Reaching Your New Digital Heights

The 4th Industrial Revolution is here, and it is the catalyst of our mindset changes as we are facing a new world of digital transformation. Mindset stands for our outlook, attitudes, and behaviors toward the world. Now that the world is rapidly changing due to technological advances, our mindset needs to leap with the trend and enable us to excel in the new digital era.

Many books may have touched on the subject of digital mindset but this book takes it to a new level. The new Cognitive Model of Digital Transformation, introduced in and followed by this book, is dedicated to digital mindset leaps from key concepts and comparative approaches to best practices.

The Cognitive Model of Digital Transformation categorizes the process of digital mindset leaps into five different layers, from Layer 1 as the foundation or starting key concepts, Layer 2 for digital ways of thinking, Layer 3 on digital behaviors and capabilities, Layer 4 on digital transformation, all the way to Layer 5 of wisdom in digital space, walking through the entire journey of digital mindset leaps.

This book intends to help get your mindset adapted and ready to navigate digital transformation along the right track. Enjoy this book and its amazing journey of digital mindset leaps.

David W. Wang is a senior business executive of the next-gen network, IT, and digital solutions, an outstanding digital transformation evangelist, and a consulting principal in the global marketplace. David is also the author of *Software Defined-WAN for the Digital Age: A Bold Transition to Next Generation Networking* (CRC Press, 2019) and *Cash in on Cloud Computing: A Conversational Guide for Buying and Selling Cloud Services* (Meghan-Kiffer Press, 2015).

Reaching Your New Digital Heights

32 Pivotal Mindset Leaps of Digital Transformation

David W. Wang

CRC Press
Taylor & Francis Group
Boca Raton London New York

CRC Press is an imprint of the
Taylor & Francis Group, an **Informa** business

First edition published 2024
by CRC Press
6000 Broken Sound Parkway NW, Suite 300, Boca Raton, FL 33487-2742

and by CRC Press
4 Park Square, Milton Park, Abingdon, Oxon, OX14 4RN

CRC Press is an imprint of Taylor & Francis Group, LLC

Library of Congress Cataloging-in-Publication Data
Names: Wang, David W., author.
Title: Reaching your new digital heights : 32 pivotal mindset leaps of digital transformation / David W. Wang.
Description: First edition. | Boca Raton : CRC Press, [2023] | Includes index.
Identifiers: LCCN 2023003468 (print) | LCCN 2023003469 (ebook) | ISBN 9781032304557 (hbk) | ISBN 9781032304564 (pbk) | ISBN 9781003305187 (ebk)
Subjects: LCSH: Industry 4.0.
Classification: LCC T59.6 .W36 2023 (print) | LCC T59.6 (ebook) | DDC 004.67/8--dc23/eng/ 20230214
LC record available at https://lccn.loc.gov/2023003468
LC ebook record available at https://lccn.loc.gov/2023003469

ISBN: 978-1-032-30455-7 (hbk)
ISBN: 978-1-032-30456-4 (pbk)
ISBN: 978-1-003-30518-7 (ebk)

DOI: 10.1201/9781003305187

Typeset in Sabon
by MPS Limited, Dehradun

This book is dedicated to my family members: Susan, Hyden, and my parents overseas. Without your support and inspiration, this book would be impossible.

Contents

Preface – The Cognitive Model of Digital Transformation

"To blur the lines of machines and humans" means we will become an integrated new entity of digital transformation. Ironically, mindset changes often fall behind technological, economic, and social-cultural changes, which is the problem this book intends to address.

We are now riding on the waves of the Fourth Industrial Revolution (4IR) that is primarily driven by digital transformation consisting of such digital technologies and applications as nanometer chips, software, virtualization, fiber cable-enabled wired and mobile networks, 5G, Internet, artificial intelligence (AI) and automation, big data analytics, and cloud computing.

Digital technologies and applications, however, are not something brand-new. The Third Industrial Revolution, which started about a half-century ago, already invented and developed personal computers (PCs) and the Internet. The distinction between the 3rd and 4th Industrial Revolutions lies between these two buzzwords: digitization and digitalization.

Digitization means we primarily use digital technologies and devices as supporting tools. Digitalization, on the other hand, has started to blur the lines between machines and humans, converging advanced digital technologies into part of our lives, complementing or even replacing what we do, and lifting our societies to the next level with advanced productivity, intelligence, and connectivity.

"To blur the lines of machines and humans" means we will become an integrated new entity of digital transformation. In the face of such significant technology-powered changes, some common questions each of us may have include: What will digital technologies bring about? What should I do in mid of all these rapid changes? How can I stay with the trend or even ahead of digital transformation?

The answer is that to excel and succeed in the digital age, we need to keep our mindset open and adapt to this exciting ramp-up of technological advances. Based on the *Harvard Business Review* published in May 2022, "A digital mindset is a set of attitudes and behaviors that enable people and organizations to see how data, algorithms, and AI open up new possibilities and to chart a path for success in a business landscape increasingly dominated by data-intensive and intelligent technologies" (source: https://hbr.org/2022/05/developing-a-digital-mindset).

Ironically, mindset changes often lag behind technological, economic, and social-cultural changes, which is the problem this book Intends to address. The good news is that just as a child goes through some mental development leaps in learning the new world, we can also achieve mindset leaps for digital transformation by following the cognitive model this book introduces.

One key note here is that when we talk about "cognitive," it means much more than just knowledge learning and accumulation. It is also about technology assessment and judgment. Because only when we gain technological insight and take the optimal approaches to adopt new technologies, the real leaps can happen.

As complex as the process of mindset adjustment and changes can be, the 32 pivotal leaps covered in this book will be elaborated on and orchestrated through the Cognitive Model of Digital Transformation. The Model, as the book's trademark illustrated in the below chart, categorizes the cognitive process of digital mindset into five different layers, from Layer 1 as the foundation or starting key concepts, all the way to Layer 5 of wisdom in digital space, walking through the entire journey of digital mindset transformation.

THE COGNITIVE MODEL OF DIGITAL TRANSFORMATION

Layer 1 Key Digital Concepts, Disruptions, and Milestones

Layer 2 Digital Thinking from the Building Pillars of Technologies

Layer 3 Smart Digital Behaviors and Capabilities – Sense, Think, Connect, Communicate, Visualize, and Act

Layer 4 Digital Solution and Application Models – Operation Transformation, Customer Experiences, and ICT Ecosystem

Layer 5 Wisdom in Digital Space – Equilibrium, Leadership, and Culture

This Cognitive Model of Digital Transformation takes shortcuts to achieve our mindset leaps as we focus on those vital digital concepts, thinking, behaviors, application models, and wisdom of management, which are holistically summarized, succinct in critical points, proved by successes, and easy to keep in mind. The more we get immersed in these pivotal practices, the quicker our mindset can kick off the leaps of adapting to the new Digital Age.

By addressing in detail the 32 leaps for digital mindset transformation, this book will effectively assist you in reaching new heights in this 4IR-defined digital era. Each of the 32 chapters in this book covers the background and concepts, innovative leaps that are taking place, new outlooks associated, and insightful case studies.

For instance, this book highlights the critical role of data, as digitalization is all from and about data, and all digital technologies are being developed around data. Our mindset, therefore, needs to adapt to the practices of data collecting and converting, data processing and analyzing for decision-making, data communicating, data storage and security, data-driven applications and automation, and so on.

This book is well suited for business and institution personnel, from executives to managers and specialists, as well as younger generations like college students involved with the practice and learning of digital transformation. Those who want to ride on the waves of digital trans-formation, prepare the right mindset to win, and reach new heights during the rapid technological-social changes, should read this book. While this is not a textbook, it can also be a good reference source for digital education.

Digital mindset leaps make the critical and unique parts of digital transformation, i.e., the ramp-up of the Fourth Industrial Revolution (4IR). Digitalization is changing our work, life, and society and merging with human thinking and activities like a brilliant partnership. Join this partnership, start the exciting journey, and make the mindset leaps happen – which will enable you to win and lead in the new Digital Age.

David W. Wang
Oak Hill, Northern Virginia,
May 2023

Acknowledgments

Special thanks to Ms. Gabriella Williams, Editor for CRC Press (Taylor & Francis), as well as her whole publishing team, as well as to Mr. Lee Wade, Founder of Exponential-E Ltd, for their professional work and strong support during this book's publication.

Appreciation to all the clients and partners who have supported my company ITCom Global, LLC, and business over all these years.

All figures in this book are used with courtesy from Pixabay.com.

Part I

Layer I – Key Digital Concepts, Disruptions, and Milestones

Layer 1 makes the foundation of the Cognitive Model of Digital Transformation introduced by this book. It lays out conceptually what digitalization is, where the 4th Industrial Revolution powered by digitization is from, and how it can significantly impact and change our lives and work. Binary systems, big data, software architecture, communications, and artificial intelligence are the fundamental DNAs of digital transformation.

> There is no alternative to digital transformation. Visionary companies will carve out new strategic options for themselves – those that don't adapt will fail.
>
> – Jeff Bezos, Founder of Amazon

> You can't delegate digital transformation for your company … You and your executives have to own it! Executives need to engage, embrace and adopt new ways of working with the latest and emerging technologies.
>
> – Barry Ross, CEO & Co-founder, Ross & Ross International

DOI: 10.1201/9781003305187-1

Chapter 1

From 0, 1, Toward Digital Infinity – The New Binary World

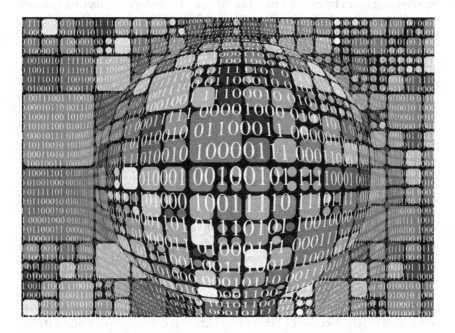

The binary world of digitalization consists of data, computing, network, end devices, and users.

We often use the word digital or its synonyms, such as digitally, numerically, computerized, digitized, and computer based. Digitizing the data is the start of digitalization. Digitization means turning regular and analog data content and format into digital format (binary systems made of 0 and 1) for better data quality, programmable storage, accessible communication, and analytical processing.

Analog means something that bears a physical analogy to something else that easily appeals to our direct observation and understanding; something that has some common traits with something else, like a mechanism, such

DOI: 10.1201/9781003305187-2

as bicycles with wheels or clocks with hands and faces. Such mechnism is analog becuase it does not break everything down into binary codes to work. For example, an analog signal of our voice record can show being strong in the electric current when we are screaming and weak when we are murmuring, just physically analog to our real voice. Computers, however, do not understand strong or weak and high or low pitches in a voice but only know how to process numbers.

What are digits? Everyone knows it is about numbers 1, 2, 3, 4, 5, ... This is called the decimal system. But to make digitalization happen, we do not use decimal numbers because they are too complicated for computers to handle. Instead, we use binary numbers for digital and computer circuits, representing them by either a number "0" or "1." Binary numbering systems work best for digital signal coding, using only 1 and 0 to form different figure combinations.

Then we can create methods called algorithms (referring to Chapters 4 and 16) to convert binary into decimal numbers. Such as, the decimal number 15 is represented as 1111 in binary, the decimal number 10 is coded as 1010 in binary, and so on.

Analog is the opposite of digital; the real world is mostly analog. In the past, we did not know how to represent, store and process information and knowledge via numerical codes, and they were isolated, spread, and wasted all over the place. Now with digitization, information and knowledge can transform and consolidate into a great virtual depository made of digits and programmable codes. This is creating a brand-new digital new world.

When you open a webpage and read the content online in the language you know of, be aware that computers initially do not process, store, and communicate in our ordinary languages. Computers natively do not understand English, German, or Chinese. Instead, we use HyperText Markup Language (HTML) to create pages on the Web. In Google Chrome, if you open any page with content, then right-click the page and select "View Page Source," it will show the program codes behind the scene: HTML source codes that work with machine codes or binary numbers that computers can understand.

The philosophy of digitalization is to attempt one fabric for all or standardize the data processing procedure and lift our productivity exponentially. One fabric doing all or many into one model means digitalization allows data of all kinds in all formats to be processed and intermingled with the same high efficiency. We can find that all achievements of digitalization own such characteristics.

THE INNOVATIVE LEAP – STANDARDIZING AND ENHANCING THE DATA VIA DIGITIZATION

Digitization essentially means taking "analog information" and encoding it via programs into 0s and 1s so that computers can store, process, and transmit such information. Hence to use a computer to process human voice signals, for instance, we need to translate the analog signals into digital codes first. This is called to be digitized.

One big event of such digital transitions happened on June 12, 2009, when all USA full-power analog TV broadcasts were required by law to end and start to be exclusively digital broadcasting. Why do we adopt digital TV? Better service quality, more robust data processing capabilities, higher resource efficiency, and cost-effectiveness are all at stake for such a digital transition.

Digital signals have distinct advantages over analog signals. First, they maintain quality better than analog because they are less susceptible to interference. Second, because digital transmission has a higher bandwidth, more information can be transferred using fewer cables or connections. Lastly, signal conversion is much cleaner when transferring digital to digital.

Such digital transitions in TV broadcasting thus can bring about many good results as explained in following details.

To the general viewers, digital TV means better sound and picture quality. A digital traffic does not suffer from the same signal degradation as an analog signal in audio and video. Anything in the past, like "vague voice and snowy TV pictures," can be mitigated and fixed over digital processing and format.

Also, a digital signal can carry more data than an analog signal or call it higher digital bandwidth, so more sound and video can transmit to the audience's TV set. For example, a TV station can transmit up to three standard-definition (SDTV) transmissions; that is three different shows on one single digital channel, such as on Channel 7, the audience now has the choice of watching 7-1, 7-2, or 7-3.

The biggest benefit of switching to a digital signal for TV broadcasters is that it will free up valuable channels of the broadcast spectrum. The extra bandwidth can be used for other purposes, such as advanced wireless and public and safety services.

NEW MINDSET FROM DATA DIGITIZATION

Technical Outlook – Analog–Digital Conversion

Telecommunication is a perfect example of analog-to-digital conversion. As briefly mentioned above, there are two general data types: analog and digital. Nature is analog, while a computer is digital. All digital data are stored as binary 0 and 1 digits.

In telecommunications, we use analog and digital signals to carry information. The difference between both signals is mainly that the analog signals have continuous electrical signals, like what we hear and see in the real world. In contrast, digital signals have non-continuous electrical signals, which are more electronic in nature. Digital transmission is applied to achieve high reliability, plus the cost of digital switching systems is much lower than that of analog systems.

However, the analog signals are still of great importance, because they make up most voice, radio, and television communications from the end users or applications. To make the whole process work together seamlessly, analog signals must be subjected to analog-to-digital conversion – using a device called analog-to-digital converter. This is then transformed as a series of voltage pulses of equal width representing binary "1" and "0" digits.

Business Outlook – A Digital Data-Centric Business

Data are not new; humans have recorded facts as symbols, numbers, and letters for thousands of years. What is new is that data are now translated into digital data and growing at an unprecedented pace. The volume of data going digital and produced daily is exploding due to the proliferation of new digital devices, services, and sensors throughout the economy and society.

The increasing volume of data creates a deep well of possibilities for scientific discovery and for improving existing or inventing new products and services. Generating insights from big data enables businesses to put customer needs first, grow new markets and opportunities, and create efficiencies that drive costs down and increase revenue.

Today companies simply cannot survive without data. Data provide great insights, such as consumer behavior or market conditions, that can reduce operational costs, track and improve performance, and optimize expenses. Data enable companies to identify opportunities, predict trends, and stay ahead of their competition. Data operate digital devices, systems, and Internet of things (IoT) to make the business more productive and customers better served.

Application Outlook – Starting from Digital Data Collection

We can feel out and experience digital data in multiple ways on a computer system or platfrom. For example, music, movies, and games can be stored as sequences of binary 0s and 1s being interpreted and performed by a computer.

Digital data collection, often called tagging, usually marks the start of a digital journey. Before digital data collection became possible, we used pen and paper to record information gathered in the field. Now the most common form of tagging online for eCommerce is to collect information about the webpage or actions taken on the webpage and then send that to a tool that sorts out that data and makes it ready for analysis.

Digital data collection can also happen via electronic technologies such as tablet computers or smartphones. Often, we see salespeople or customer service reps today just key in data as they go. Collecting ongoing information can also save businesses money by building a database of customers for future marketing and retargeting efforts.

For instance, Radio Frequency Identification (RFID) is one method of automatic identification and data capture on merchandise. It is a technology whereby an RFID reader captures digital data encoded in tags or smart labels of a merchandise via radio waves. The RFID reader, a network-connected device, uses radio waves to send signals that activate the tag. Once activated, the tag transmits a signal wave about the merchandise back to the antenna, where it is translated into data, usually as an identifying inventory number.

INSIGHTFUL PRACTICE

The Kodak's Lessons Learned

This is a digital camera that can take pictures with instant images and no need of films.

Businesses and organizations nowadays need to go digital quickly. Kodak is one of the painful lessons learned in this regard. Still, remember Kodak – the one-time giant in the photograph industry? Kodak was founded in the late 1880s, reached its peak in the 1970s but filed for bankruptcy in 2012. Kodak was at the forefront of photography for almost an entire century with many innovations and inventions, making this art and entertainment accessible to ordinary consumers.

But why did Kodak eventually fail? The simple answer is that it did not go digitally big and fast enough. The company did invent its "first digital camera" back in 1975 but then almost paused there. Arrogant and blind-minded, the company refused to take advantage of its market-leading position to pivot to digitalization. Leaders in Kodak remained blindly confident that the firm's marketing power could persuade its customers to ignore digital photography.

Kodak believed that the new digital technology would erode its film business – a typically short-sighted mindset when incumbent leading companies face and adapt to innovations. Unfortunately, as digital technology quickly grew, more and more consumers did not seem excited to pay for photo printers anymore, and Kodak was headed for its continuous loss for over seven years. Eventually, Kodak failed because it did not capture what its customers were buying and missed the driving force of new digital technologies.

The difference between film and digital photo taking is mainly about the media used. The traditional camera uses film for capturing and storing pictures. On the other hand, in digital photography a digital lens captures the image and stores it in flash memory. The benefits of digital photography, as most of us know it today, are revolutionary, huge, and self-evident, as summarized in the following points:

- Film is expensive to buy and process. Digital photos offer immediate satisfaction for the best choices. No film is needed.
- Massive storage (aka electronic album) space for photos, even including video cameras.
- Multiple and smart functions available; easy to edit, display, and share.
- Smaller and lighter cameras for portability, allowing home printing.
- Point-and-shoot technology is easier to learn and operate for satisfactory photos.

If a company could not offer the above camera and photo benefits or the new value chain to the market and consumers, they would become outdated. Thus, when Kodak decided to cling to analog cameras for far too long, it lost the digital battle to its bankruptcy.

The Magic of Voice over IP

Digital services are typically related to media and communication. It generally, if not always, can lower costs than the traditional delivery method. We have noticed that phone calls nowadays cost much less whether we make local, national, or international calls. Also, we can make calls from regular phones, smartphones, PCs, and tablets. The magic maker behind the scene is voice over IP (VoIP).

VoIP converts the voice analog signal into a digital signal, compresses it into packets, and sends it over the Internet. Hence basically, voice is transmitted as data. A VoIP service provider transmits the call over the data packet network. The data signal is uncompressed on the receiving end and converted back into the analog sound you hear through your handset.

Traditionally, the Telco central office (CO) transmits local analog signals and later converts the analog voice signal to digital format. The traffic is transmitted over the public switched telephone network (PSTN), which is less efficient than the Internet network. VoIP allows us to connect an analog phone to the interface of an Internet router; then the phone sends an analog voice signal to the router, and the router converts the analog signal to a digital signal. This makes VoIP easier and more popular because the Internet is more available than PSTN and a lot cheaper.

How much cheaper? It can cost up to 2/3 less than traditional phone services, eliminating cost concerns over local, long-distance, and international calls because the Internet is a global data network, and when we route both voice and data over the Internet, it becomes pretty cost-effective. The digital format and Internet make VoIP possible, with such advantages as easier accessibility – calls can be made via phones and PCs; clearer voice quality and carrier-class reliability; supporting multitasking or unified communications in today's office environment.

Digital Data Links and Modern Wars

In military terms, the generation gap between the 2nd and 3rd generation weaponry lies in the weapon system's digital information and communication capabilities. The 2nd generation weapons are mechanical and still require manual operation from humans. In contrast, the 3rd and above-generation weapons started to use computers and data links to operate more quickly, accurately, and powerfully. The outcome – the blind and deaf 2nd generation weapons would lose the battle.

A data link essentially enables two parties to communicate messages. In ancient times, for example, smoke signals from the top of a hill or mountain for something urgent and critical were used by people to convey messages across a distance. The parties involved would agree on the signal indication or context in advance, such as the enemy starting a surprise attack, an essential spot in danger, etc.

In modern warfare, such critical messages become digital, and digital data links play a crucial role in joint military operations. The US military has developed the Tactical Common Data Link (TCDL) to send secure data and streaming video links from airborne platforms like recon planes or drones to ground stations like tanks and other military vehicles. This is the so called Digital War – technologies and media are transforming how wars are fought, experienced, represented, and conceptualized.

Digital warefare technologies would enable ground troops to know precisely what is happening during an operation – they can make timely and crucial decisions and establish information superiority on the battlefield. For the air force, the fighter data link provides the aircrew with immediate access to unprecedented amounts of critical intelligence. Such information includes friendly, unknown, and enemy flight data and essential data elements, such as location, direction of flight, altitude, and aircraft type. Today the smarter side will win the war, and digitalization is the smart enabler.

Chapter 2

From 1st to 4th Industrial Revolution (4IR) – Exponential Changes

The 4th Industrial Revolution is about digitalization, artificial intelligence, and automation.

People may wonder why we are now in this digital transformation and what exactly it is. Where is it from? Where does it head to? A simple high-level answer is: because we are in the 4th Industrial Revolution (4IR), which comes from the 1st Industrial Revolution starting in the 18th century, has changed human lives via technology and production advances, and is heading for an even greater future of technology-powered innovations.

Talking about Industrial Revolution, it is crucial to understand these two words: Industrial and Revolution. Industrial has two implications. First, it is not the agricultural way of production anymore; second, it involves manufacturing, mass production, and standardization. Revolution

DOI: 10.1201/9781003305187-3

means it is not about evolution; instead, it is about disruptive technological inventions, work and living model breakthroughs, and brand-new social-cultural norms.

The 1st Industrial Revolution started around 1765 when steam power started being used in textile and other machinery manufacturing industries and transport like trains and ships. For example, before the mid-19th century and the Industrial Revolution, textiles were manually made or handspun, and the material was primarily from wool. Then with the invention of the spinning wheel and the loom powered by the new steam engines, cotton was produced quicker and eventually replaced wool in the textile field.

The 2nd Industrial Revolution emerged in 1870 when electricity was invented and generated. Factories also started to use assembly lines for mass production. All kinds of electricity-powered machines were invented and manufactured, including telegram, telephones, lights in people's homes, the radio, TV, movies, the refrigerator, and microwave ovens, to name a few.

The 3rd Industrial Revolution began in 1969 when businesses and the government started to use computers and the Internet. And then came personal computers (PCs), the worldwide web, and smartphones. With the aid of computers and computer programs, we could automate an entire production process – without human assistance; a small case like MS Office software we use to enhance our daily office work. Large to the scale of factory production line automation, examples of this are basic robots that perform programmed sequences without human intervention.

THE INNOVATIVE LEAP – THE 4TH INDUSTRIAL REVOLUTION (4IR)

Then the 4IR kicked off around 2011 when advanced digital application technologies lifted manufacturing, transport, healthcare, and other verticals to the next level and carried implications for the future of industrial development. The 4IR was primarily led by digital transformation consisting of digital technological developments in high-speed mobile Internet like 5G and 6G, artificial intelligence (AI) and automation, big data analytics, and cloud technology.

As highlighted in the Preface of this book, it is worth pointing out that digital technologies are not something brand new, and the 3rd Industrial Revolution has come up with PCs and the Internet. The distinction can be highlighted between these two buzzwords: digitization and digitalization. Digitization means we use digital technologies and devices primarily as tools. Digitalization has started to blur the lines between machines and humans and make digital technologies part of our lives, complementing or even replacing what we do, and advancing our societies to the next level with more productivity, intelligence, and connectivity.

A good example is our use of smartphones nowadays. Before smartphones, we also used cell phones but mainly as a talking tool with little personal attachment. But now, without the smartphone handy, many of us can hardly function in our daily lives and even feel like losing a real partner. Why is that? Because smartphone makes a good pioneer application of the 4IR, which is becoming an indispensable part of our lives and work by all means.

4IR is characterized by applying information and communication technologies to different industry verticals. The results of the 3rd Industrial Revolution are now expanded by a network connection and a digital reality on the Internet. Networking all systems leads to "cyber-physical production .systems" such as smart factories, smart offices, smart cities, and smart homes. Production systems, components, and people communicate via a network, and the working process is getting autonomous.

NEW MINDSET FROM 4IR

Technical Outlook – Toward AI-powered Automation

4IR bears its trademark of blurring boundaries between the physical, digital, and biological worlds. It is a fusion and ecosystem of advances in AI, cloud computing, the Internet of Things (IoT), 5G, genetic engineering, quantum computing, and other technologies.

What is 4IR targeting to accomplish? It marks a new era of innovation in technology which will enhance human–machine partnerships, unlock new business opportunities, and fuel economic growth across the globe. 4IR will reduce transportation and communication costs across nations, making global sourcing of raw materials and other trade possible.

AI-powered automation, such as machine learning and deep learning, combines task and process automation with intelligent analysis to accelerate innovation and digital transformation. The synergy of robotics and AI technologies makes Intelligent Automation (IA) possible, and they jointly empower rapid end-to-end business process automation and drive digital transformation.

Business Outlook – Hybrid Human and Machine Work

We are underway with digital transformation. Since digital transformation is more than technological breakthroughs and brings about innovations across the board in our ways of living, business models, and social development, our mindset must follow through and immerse ourselves in digitalization.

Creating smarter, more accurate AI systems requires a hybrid human–machine approach. Smart machines are helping humans expand our abilities in three ways: amplifying our cognitive capabilities, interacting with customers and employees to free up the workload for core tasks, and embodying human skills to extend our physical abilities.

The automation of jobs has boosted productivity in organizations. From procurement to sales, the supply chain to telecommunications, 4IR has helped reduce costs and improve the bottom line. Employees have been liberated from manual tasks to concentrate on higher problem-solving skills. Manufacturers are using AI-backed analytics and data to reduce unplanned downtime and enhance efficiency, product quality, and the safety of employees.

Application Outlook – 4IR Reshaping Our Society

As we cope with the COVID-19 pandemic, 4IR technologies play a pivotal role in maintaining productivity. It is easier to imagine how the world can handle COVID-19 effectively with the emerging 4IR technologies in place. We see that the IoT, AI, big data analysis, blockchain, and intelligent manufacturing systems are rapidly mainstreaming to enable faster digital transformation and delivery of solution-oriented services. For example, AI/ML-defined electric vehicles (EVs) are changing our driving habits, car industry, and cultural ecosystem.

Based on the Boston Consulting Group (source: https://www.bcg.com/), 4IR or Industry 4.0 refers to the convergence and application of nine digital industrial technologies: advanced robotics, additive manufacturing, augmented reality, simulation, horizontal/vertical integration, Industrial Internet, the cloud, cybersecurity, and big data and analytics.

4IR is reshaping and reforming government, education, healthcare, and commerce – almost every aspect of society. Meanwhile, 4IR is about more than just technology-driven change; it is an opportunity for everyone to excel, including leaders, policy-makers, and people from all nations and walks of life. People can harness converging technologies to create an inclusive, human-centered future. Digital technologies can also change our values, like the things we value from traditions and how we value them now. From productivity standpoint for example, an office word-processing task that used to take a week or so to be completed, now can be expected for its completion in a few hours via the digital way.

INSIGHTFUL PRACTICE

The World Became Smaller with Trains and Cars

Back in the mid-18th century, the invention of the railroad and the steam-powered locomotive from the 1st Industrial Revolution opened up a whole new world in transportation. Trains could travel wherever tracks go as they were built and laid out. Long-range transit was no longer limited to rivers and canals.

Massive transport changes took place in the 1700s and 1800s. For a long time, canals provided the means of transporting many goods at once via waterways. One journey on the canals than by road can transport a larger load of goods. Then came the most significant transport development in the 1800s, the invention of the railway. For example, the deployment of the railways in Europe drastically changed time and distance both in concept and reality during the Industrial Revolution. Before using railways, people travel had to use other old means of transportation, such as walking and riding horses.

The creation of power machines and factories also brought many new job opportunities and the migration and transiting needs of people from the countryside to cities. Transport became a reality for both goods and people. Then individual and family transport changed entirely when the motor car got invented in the 1890s. As a result, road and railway transport in both the European and American continents had taken over sea transport.

Historians and economists agree that any society entering the industrial age needs to have an effective transport network. People's standards of living around the world fast increased because, for the very first time, world trade became possible, namely, goods could be shipped and delivered across the globe. They promoted trades for other products. Modern transportation was also essential to the nation's industrialization. Busy transport links increased the growth of cities. The transportation system thus helped to build a new industrial economy nationwide.

Electricity Lightening up the World

Electricity is the greatest invention ever in human history because it opened our lives to a whole new civilization. The father of electricity, Michael Faraday, from the 19th century, was a great English scientist and the champion for discovering electromagnetic induction, electrolysis, and dia-magnetism. In 1882, the practice of electricity generation at central power stations got started. It happened in New York City, where a steam engine-driven dynamo at Pearl Street Station produced a current that powered public lighting in the neighborhood.

Electricity became important during the 2nd Industrial Revolution in the late 1800s, which began through the discovery of electricity and assembly line production. Before it entered homes, electricity started to power many areas, such as industry, railway transport, and street lighting. Electric lights, for example, stretched factories' operation time and produced more goods. The great Thomas Edison inventions – electric lighting, the telegraph, and the telephone – lifted our lives to the next level.

The 2nd Industrial Revolution invented electricity and bulb and enlightened human civilization.

Since electricity was invented, it also became the foundation of many new inventions. It adds light to the human world and is a pivotal solution to most modern problems. Electricity not only powered factories but also got used in transportation to power boats and trains. Electric power began displacing steam power as the primary power resource for industries. The creation of many media industries followed through, including telecommunications and the radio industry.

To date, electricity provides clean, safe light around the clock and globe, empowering the new digital world we link with our smartphones and computers. The energy revolution also leads to emerging new markets, businesses, and job openings. We cannot live without light; a business cannot grow without power. When electricity came into being, it lightened up a new civilization.

Computers Changing the Modern Society

The 3rd Industrial Revolution shifted from mechanical and analog technology to digital electronics. The new revolution began in the latter 20th century when we adopted the widespread use of digital computers, data-keeping, and many other innovations.

A computer – an electronic device that stores, retrieves, analyzes, calculates, and processes data – is an extension or complement of human intelligence and capabilities. Software or applications must function on top of a computer; regardless it is a physical or virtual computer. In our time many daily activities take place as online services and products. This can only be possible via computers.

Computers began to enter the mainstream in the 1960s, transforming businesses worldwide. Then during the 1980s, PCs came into being for consumers and individuals. The Internet and the 3rd Industrial Revolution could only have flourished on top of computer-wide adoption. Computer technology has increased the speed of information sharing, and we can draw conclusions faster and use the knowledge to get more effectively.

The computer makes a powerful symbol of digitalization, and the world has become faster and smaller with digitally powered computers. To date, computers benefit the business and personal world from all aspects of life with better efficiency: buying and selling products, increasing workers' productivity, communicating throughout the world, enhancing our knowledge, job influences, entertainment, research, and paying bills. Computers help to improve healthcare, finance, defense, education, robotics, big data and analysis, and more.

Chapter 3

From Data to Big Data – Oil of the Digital Age

Big data analytics includes data collection, verification, processing, reporting and decision-making, and responding.

Data are the energy source and blood flow for digitalization to materialize and grow. Digitalization is all about gathering, processing, and storing data and then communicating and applying information extracted from data. Without data sources, digitalization would have nothing to process and produce from, thus becoming baseless and impossible. In addition, good data allow organizations to measure performance and productivity to establish baselines, benchmarks, and goals to keep businesses moving forward.

DOI: 10.1201/9781003305187-4

This is like we use a smartphone which is a personal digital device and platform. Our smartphone will function well when it can receive and send data (e.g., taking phone calls, surfing online, receiving emails, and messages), collect data (e.g., taking photos, making notes, scanning codes, etc.) from outside, and use data (e.g., via all kinds of apps such as Google Map, Uber, Whatsapp, etc.).

Worst case, say we are on an airplane and have to put the smartphone onto airplane mode, cutting off the device's link with the outside. But we may still use the phone's calculator, play some games, or do some reading. Remember that we still need to input data to the calculator, and the games or documents need to be already downloaded (input) to our phone's files.

The critical point is that digitalization is all about data. Without data, digitalization cannot happen or function. Nowadays, we often run into a popular buzzword, "Big Data." What does big data mean? Big data is about large, more complex data sets, hard-to-manage volumes of data – both structured and unstructured – especially from new data sources. Organizations and businesses are immersed with big data daily in the digital era.

Structured means the data come in or are gathered in a fixed format, like via a standard or online form. Why do most businesses use Microsoft Excel or Access tools to store and process daily data? Because both Excel and Access have data records in a fixed format, making data processing, and storage much easier to handle.

Unstructured data come in a random format or no format at all. Now businesses know that unstructured data can be very valuable, such as via the client's purchasing receipts, customer service phone or chat records, and service requests or online searching requests. Therefore, new algorisms and software tools are being developed to organize and process unstructured data.

THE INNOVATIVE LEAP – DATA IS POWER

Ultimately, it is about how organizations make use of the data that matters. Big data can be analyzed for information and help us gain insights so as to make better decisions and strategic business moves. This is to say that data must become information and insight to produce values.

Some data scientists at IBM break big data into four dimensions: volume, variety, velocity, and veracity. While we have mentioned volume and variety above, velocity is about the speed that data can be processed to generate information. Sometimes this is hard to achieve due to the large volume and wide variety of data. But if it takes too long to develop information from the data, the time value of the information may get lost, and good opportunities can be missed.

Veracity is about the truthfulness or accuracy of the data. This may become hard to manage, too, especially from unstructured data sources. Hence while big data makes great sources, there are two major challenges: How to convert them into useful intelligence quickly and accurately? How to take advantage of big data and use it wisely for business benefits and growth?

All these successful technology giants today, such as Amazon, Facebook, Apple, Tesla, Google, Netflix, and so on, use or even rely on big data for their growth and development. Why is Tesla more successful in developing its driverless cars? Because it has gathered much more data from various road conditions worldwide to make its car computers smarter and safer for self-driving.

NEW MINDSET FROM BIG DATA

Technical Outlook – Data Analytics

First, we need to understand better our data sources, such as media, eCommerce web portals, cloud platforms, and IoT. Media is the richest big data source, with valuable insights into consumer preferences and changing trends. The eCommerce web portal generates big data that is widespread and easily accessible. Cloud storage also accommodates structured and unstructured data and provides businesses with real-time information and on-demand insights. IoT makes another valuable big data source, usually generated from the sensors connected to electronic devices.

Businesses today can use an amalgamation of traditional and modern databases to acquire relevant big data. These databases can extract insights via advanced analytics, machine learning, and other AI technologies to drive business profits. More diverse data drives advances in processing power and the rise of edge computing.

Qualitative and quantitative data analysis techniques are mainly used for data analysis. Qualitative analysis involves identifying, examining, and interpreting patterns and themes in textual data and drawing intelligent judgments and conclusions. Quantitative data analysis refers to the use of computational and statistical methods. This approach focuses on dataset statistical, mathematical, or numerical analysis. These two data analysis techniques can be used independently or in combination with the other, which helps business leaders and decision-makers gain business insights from different data types.

Business Outlook – Data-First Strategy

An organization today will need a data strategy which is a plan designed to improve all of the ways data are being acquired, stored, managed, shared, and used. These are the five core components of a data strategy – identity, store, provision, process, and govern. They work together as building blocks to comprehensively support data management across an organization.

For the most successful businesses, rather than setting goals based on gut instinct and observation, examine how data enters and moves through their organization in granular detail. By understanding their data workflows, businesses can quickly identify the areas of inefficiency, governance weaknesses, security gaps, and where money is being wasted. This is called the Data-First Strategy.

Once businesses have identified the issues, such as storage needs, lack of processing tools, security concerns, and so on, they can consider adding technology to address them. With a data-centric strategy, improvements are real-world intelligence based versus adopting fancy new technologies that are expensive and hardly a fit for the business needs.

Application Outlook – Capturing the Values of Data

Big data by word refers to a large amount of data; more important is the information extracted from big data by analyzing the vast amount of complex data sets. Big data is both now and the future. Businesses and organizations in the digital era will continue to be shaped by evolutions in how we store, move, and understand data.

Traditional data analysis occurs incrementally and reactively. The flow is like this: An event occurs first, then data are generated, and the analysis of this data follows. Such an approach can help businesses learn the impacts of given strategies or changes in a post-mortem mode on a limited range of metrics over a specific period. Now big data analysis can occur in real-time and trigger quick actions. Such a real-time approach offers businesses a more dynamic and holistic understanding of their needs and strategies.

Because of this real-time characteristic of big data, enterprises and organizations use big data analytics to optimize their business intelligence and analytics quality. Big data can move past slow reporting tools dependent on reactive analysis technology to more intelligent, responsive applications. This leap enables greater visibility into customer behavior, business processes, and overall operations.

INSIGHTFUL PRACTICE

Tesla – Big Data Powering the Car

What energy does a Tesla vehicle use? You may say electricity. That answer, however, is only half true. Tesla also uses virtual and more powerful energy called data: it is a "data-oriented" and "data-driven" company. Data are the primary reason for Tesla's success since data collection and analysis are at the center of everything this firm does: design, manufacturing, customer data, roads, satellite, and even discussion boards. Such a data-first strategy generates new insights to maintain Tesla's leading position in autonomous cars and the electric vehicle (EV) market.

Unlike traditional car makers who are all about mechanical designs, Tesla is all about its collected more than 1.3 billion miles of data in a real-world scenario. With such big data, it constantly trains its auto-pilot system using a deep neural network – a machine learning model. The fleet learning technology developed by Tesla is a core capability with effective updates delivered to every product.

Using real-time and predictive analytics, Tesla can also enhance better customer service and car maintenance. The models can predict car maintenance needs and use smart alerts to notify the driver when it is time to bring the car in for repairs. In addition, these systems could also automatically make recommendations for in-car adjustments based on individual driver data.

This is a Tesla electric car with big data collection and analytics capabilities to aid its auto-pilot driving.

Most traditional EV makers rely on battery technology as niche, but Tesla broadens its niches also into big data, artificial intelligence, and the Internet of things to outsmart its competitors. This brings about a very competitive bundle for Tesla: advanced battery, software, and solar panel technology, plus its current catalog of data-driven services; this boosts Tesla's total product profitability and continuous leading role in the market.

The Amazon Big Data Success

Most of us use Amazon.com, a big data-driven company with great success. Big data have helped lift Amazon to the top of the competitive ecommerce marketplace. Amazon takes on a broker or middleman like business model. The company does not manufacture anything but links with manufacturers and tracks its inventory to ensure orders are quickly fulfilled to Amazon's customers. Big data play a pivotal role in this process; for example, it can map and match warehouse locations with customer tracks. This enables to deliver the goods to the customer's hands faster and reduces shipping costs by 10–40%.

On the customer-facing and service front, Amazon leverages its data via its recommendation engine; most of us should have experienced that. Via its web portal, Amazon collects data regarding each of its customers. On top of the data about goods the customer has purchased, Amazon can also collect and track what items were viewed, the shipping address of the users, and the reviews left by the user.

Big data is also the energy for Amazon's AI applications to take off in enhancing its customer experience and internal best practices. Amazon's AI capabilities range from predicting the number of customers willing to buy a new product to running a cashier-less grocery store. AI can also provide customized recommendations to its customers.

For instance, we often see phrases like: "Customers Who Bought This Item Also Bought ..." on Amazon sites. This is based on the rankings of choices and clicks made on the site processed and indicated by the Amezon's AI tool by following the respective tastes of users.

Google – YouTube Sensation Backed by Big Data

Google is undoubtedly a company that famously uses big data due to its special advantages in search engines. In addition to the billions of daily search requests on its search engine, Google also has access to user information through its Chrome browser and Gmail services.

Google's big data are built on the Google Cloud Platform (GCP), which offers computing, storage, databases, networking, machine learning, management, development, and security tools. It is estimated that Google processes approximately 63,000 search queries every second, translating to 5.6 billion daily searches (source: https://blog.hubspot.com/marketing/google-search-statistics).

Google has another crown jewel – its streaming service or YouTube, one of the best engines producing massive amounts of data in a brief period. YouTube is an online video-sharing service via which users can watch, share, comment, and upload their own or other videos. YouTube can be accessed on PCs, laptops, tablets, and mobile phones. It is estimated that over 1 billion hours of YouTube are watched globally daily.

YouTube data are generally unstructured, and big data analytics makes the best fit for processing such data and helps YouTube generate valuable insights. A user's activity information that YouTube collects may include such records as search terms, videos they watch, people in communication or sharing content, and Chrome browsing history they have synched with their Google Account.

Then YouTube uses big data to refine its marketing campaigns and techniques like recommending sound clips to the right audience. It also uses big data in machine learning projects to train machines, predictive modeling, and other advanced analytics applications.

Chapter 4

From Hardware to Software-Defined Architecture – The Digital Way to Go

Software is opposite to hardware and series of programs developed to operate the hardware.

Software is the opposite of hardware and includes programmed instructions, data, or codes to operate computers and execute specific digital tasks. It refers to the non-physical aspects of a computer. Software used to be deemed the magic part to make a device work the way we want it to, with hardware as the platform. It is like hardware is the stage, and software puts up a show or performance on the scene. Hardware and software tended to do things in parallel and were equally important until recently.

DOI: 10.1201/9781003305187-5

Based on the notion above, each software piece comes together with a hardware piece to make an IT machine (or called a device box), such as a server, a router, a switch, a firewall, or a personal computer (PC). As the result of IT operations growing with the applications and users, enterprises may find themselves surrounded by too many boxes.

In the virtualization chapter (Chapter 14), we mention that too many boxes would consume many resources in terms of energy, space, management, maintenance, and upgrade and that is why we started to use "virtual machines" (a software instance) to replace physical boxes to handle this "too many boxes" situation.

But only "virtual machines" are not good enough because they remain individual and point solutions and are difficult to manage. We also need a more robust centralized system to better manage virtual machines. For example, how does a firm work, troubleshoot, and upgrade its network infrastructure effectively and quickly if it owns a global data network made of thousands of switches, routers, firewalls, electric-optical converters, and other servers?

The answer is to separate hardware and software; lots of such efforts have been underway for some time. At the start, software was written for specific computers and sold with the hardware it ran on. In the 1980s, the software began selling on floppy disks, CDs, and DVDs. Before they are separated as we can do now (called decoupling, like software as a service – SaaS from the cloud), if we wanted to know how a box (e.g., a web server) was working, we still needed to access and check each box manually.

THE INNOVATIVE LEAP – SOFTWARE-DEFINED ARCHITECTURE (SDA)

Then a new idea came up: we could use a centralized software system to better control distributed hardware or network instances. We could put software in a leading control position and manage the whole network with the centralized control system.

A key technology to make this happen is modulating the software or decoupling the control plane and data plane within the software layers. The data plane (sometimes known as the user plane, forwarding plane, or carrier plane) is the part of software operating a network that carries user traffic. The control plane and management plane serve and guide the data plane and are the part that can be decoupled and centralized.

Once they are decoupled, each box function can now be represented and controlled by the centralized software. The task of managing each box can be accomplished via collectively addressing the central control software, making centralized control and orchestration possible.

Now since we can single out the software of each box that can be easily accessed and managed over a centralized platform, we can achieve a holistic better control of the whole IT OS, such as a data center or a network. This is called SDN (software-defined network) if applied to a network. If applying to a data center, this is called SD-DC (software-defined data center). Such an approach uses software-based controllers or application programming interfaces (APIs) to communicate with underlying hardware infrastructure and direct traffic on a network or within a DC.

Using our smartphone (as a hardware device) as an example, once we have a physical smartphone handy, like iPhone 8, we keep the same phone set as long as we do not change to an iPhone 11. How does Apple upgrade and patch the operating system (OS) and other software of the iPhone 8 we have handy? The answer is from the cloud via the communication network, targeting the software only. With the centralized software-defined architecture, Apple can upgrade thousands of iPhones simultaneously.

Some people may take this process for granted. But an experiment you can do is to cut off the network your smartphone has (namely, making it offnet) for three months; then, you can find out your phone is falling behind in performance and security upgrades. Because Apple cannot access your phone's software and upgrade it over the network anymore during your phone's shutdown. In a nutshell, your smartphone cannot function well without the support from the software-defined architecture.

NEW MINDSET FROM SOFTWARE-DEFINED ARCHITECTURE

Technical Outlook – Getting Software Centric

Software is the program to operate the computer system and governs the hardware components. The software production process has a set of activities involved, which is called a software process. The software industry and products can be categorized into four major areas: programming, system, open source, and SaaS.

Programming service is a sector that has historically been the most significant and includes names such as Microsoft, Oracle, SAP, and so on. These companies often develop solutions for businesses to analyze, store, and organize data or provide programs to run machinery.

System services proliferated and exploded in the 1980s with the rise of PCs. By name, we know system software meets the need for an encompassing OS, such as Microsoft's original disk OS (DOS) and now Apple's OS for iPhone.

In contrast to proprietary software, Open Source refers to a code base that was created and is free to acquire. Any developer can access, view, and revise open-source software. A developer might spot and correct errors or omissions that a program's original authors might have missed.

In recent years, Open Source is getting popular since computer programming or software development has become a vast in-demand profession with the growth of the Internet and cloud systems. Businesses are willing to join the most cost and time-effective open-source environments, such as the Linux OS. Another well-known open-source code base is the Android OS for smartphones like Samsung Galaxy and Google Pixel.

SaaS delivers its service via the Internet and is a cloud computing-based product. SaaS has become more popular than system software for business-specific needs. The software is kept on the developers' servers in the cloud, and clients access the software through the Internet. All upgrades, patches, and software issues are handled on the developer side with a subscription-based model for the client.

Business Outlook – Software-Defined Services

Microsoft claims: "Every company is now a software company" (Satya Nadella, Chief Executive Officer in 2019).

Externally, businesses must realize software now as a differentiator – it can help them distinguish from competitors and get more competitive. Organizations can use software development to improve the client's experiences, make solutions and platforms safer, more productive and efficient, and bring more feature-rich and innovative products to market.

Internally, the software is about automation and workload reduction of the organization's tasks, as well as big data analytics reporting the progress or lags in the organization's activities. This improves the efficiency and effectiveness of the company's activities. Other benefits of software systems include speed and accuracy: data can be accessed and retrieved more quickly and with greater accuracy. This can improve the productivity of employees as well as the movement of goods and the supply of goods to customers.

For example, Customer Relationship Management (CRM) software automates the process of storing new leads, capturing existing ones, and tracking future sales prospects. With the help of CRM, the sales team can quickly contact customers and keep the sales momentum going until the deal closing.

Application Outlook – Better Apps for Bigger Market

Nowadays, the application is another name for software, specifically referring to those software apps as system interfaces and users' utilities. Businesses use the software as tools for project management, data analysis, customer services, and financial services; or use software for communication as chats, video calls, and conference platforms as needed. Modular software apps can be either off the shelf or customized.

Some examples of software applications we are familiar with include web browsers, word processors, multimedia software, spreadsheet software, email clients, graphics software, etc.

Businesses are also utilizing system software on all major fronts, with the following examples. At the first impression, you may think the following are web services for security, sensors for IoT, and headsets for virtual reality (VR) and augmented reality (AR), but behind the scene, these are all run by software.

- Blockchain for safe cash distribution
- Virtual reality (VR) and augmented reality (AR)
- Cloud computing (IaaS, PaaS, and SaaS)
- Internet of things (IoT)
- Big data analytics
- Artificial intelligence
- Cybersecurity

INSIGHTFUL PRACTICE

OS as Smartphone's Brain

Smartphones have become a standard "technology partner" for many of us in today's society. These smartphones run on special software called OS designed for specific models. Some examples of this software include Android® and Apple iOS® technology. An OS is a software interface that manages and operates hardware units, e.g., a smartphone, and assists the user in using those units. In other words, OS is the software that enables mobile devices such as phones, tablets, Apple watch, and so on to run applications and other programs.

Most mobile OSs only work on specific hardware. For example, an iPhone runs on iOS, and a Google Pixel runs on Android. Our PC has OS as well, such as Windows 10, and such an OS has three main functions: first, it manages the computer's resources, including the central processing unit, RAM, disk drives, and printers; second, establishes a user interface, and third, executes and provides services for applications software.

Apple's OS is a closed and proprietary ecosystem made for tighter integration, and that is why iPhones have robust optimization between hardware and software. Generally speaking, iOS devices run faster and smoother than most Android phones while the prices remain at comparable ranges. On the other hand, Android is a more open and cost effective system. Users in Android can customize the coding of their phones with ease. Android software is available for many smartphone manufacturers such as Samsung and Google.

Our Navigation Assistants – Google and Apple Map

Many of us might be old enough to remember the old days when we took a big map book on the way of long distances driving to visit some new places, and the navigating process was pretty manual and cumbersome. Then digital mapping apps like Google Maps have come up, innovatively changed our ways of navigating, and become part of modern life ever since. Google Maps can efficiently get us from here to there, pinpointing the locations of the Uber service we are calling.

Google and Apple Maps use the GPS location on our smartphones to determine where in the world we are via the Dijkstra's algorithm by Dijkstra, who was an engineer working on the shortest path algorithm successfully back in the 1950s. Dijkstra's algorithm has made digital mapping and navigation possible. The gist of this algorithm is what powers the navigating functionality in Google and Apple Maps today.

Digital mapping (also called digital cartography) is the process by which data collection is compiled and formatted into a virtual image. The data source and the vast collection of information are gathered from businesses, websites, users, and more. The primary function is to produce maps that accurately represent a particular area, detailing primary road directions and other points of interest. Digital maps are dynamic, cost-effective, being updated in real-time. Such maps in use can represent all features and information of a given area over a single display.

Apple Map is a software using GPS and big data to offer navigation services.

Artificial intelligence has also transformed modern navigation (also referring to Chapter 7). Using historical location data and recent search queries, AI enables digital maps such as Google and Apple Map to anticipate our destination and help us navigate there using the quickest routes possible. By July 2021, Google Maps was currently being used by around 118 million adults in the United States, while Apple Maps has somewhere between 62 and 66 million user (source: https://www. justinobeirne.com/).

API – Device-to-Server Broker

It is a type of software called middleware function that allows applications to access data and interact with other software components, OSs, or microservices. Broker is the keyword here. To simplify, an API delivers a user response and sends the system's response back to a user.

Pay attention: API is not between you and the smartphone application, say the Uber app. It is between your smartphone (Device A) and the Uber service server (Device B). When you use the Uber app on your mobile phone to request a ride, the app connects online via API and sends data to an Uber server in the cloud. The server then retrieves the data you request, interprets it, performs the necessary actions (like finding a nearby driver), and sends it back to your phone.

APIs unlock a door to external software (or web-based data) in a controlled and safe way for any program. Programmed instructions or code can then be entered that sends requests to the receiving software, and data can be returned. The API acts as a broker between any two machines or software apps that want to connect and engage for a specified task. These types of web APIs are the most common but limited just to the web.

Each time you use an app like Twitter, send a test message or check the weather forecast on your phone, you are using an API which means that the app can function between your phone and some specific app servers behind the scene. Is API the same as web apps? That is a good question. The short answer is that all web services are APIs, but not all APIs are web services.

API's go-between role allows developers to build new programmatic interactions between the various daily applications people and businesses use. A programmer can use APIs to deliver solid solutions safely and rapidly and bring in applications together to perform a designed function built around sharing data and following pre-defined processes.

Chapter 5

From Wireline to Wireless Communications – Ubiquitous Connectivity

Wireless is opposite to wireline communication and can send data, voice, and video over the air.

Let's ask a telecom professional what the difference is between wireline and wireless communication services. The answer might be that wireline means people follow devices, while wireless means devices follow people. This is what mobile technology is all about: going where the user goes.

Wireline communication (also called wired communication) transmits information over a physical circuit, while alternatively, communication technologies that transmit data over the air (OTA) are called wireless or mobile communications. Mobile technologies consist of portable two-way communications devices such as mobile phones, computing devices for processing and storage, and networking technology that connects all

DOI: 10.1201/9781003305187-6

devices. In our daily lives, mobile technology is Internet enabled, including devices such as smartphones, tablets, and watches.

The discovery of radio waves in 1880 kicked off the mobile technology era, which soon led to the invention of telegraph and mobile phones. Wireless technology includes communications using more extended wavelength radio frequency (RF) and shorter infrared (IR) waves. Wireless technology can provide many values, changes, and benefits to computing and communications. It enables faster response to queries, reduced time spent on paperwork, just-in-time and real-time control, and tighter communications between clients and hosts.

RF waves are from 3 kHz to 300 GHz on the frequency scale. IR waves range from 300 GHz to 400 THz in scale. Critical applications of RF fields with 100 kHz to 300 GHz include mobile phones, cordless phones, local wireless networks, and radio transmission towers. RF is also used by medical scanners, radar systems, and microwave ovens.

The technology generation of cellular communications networks is labeled with numbers: 1G, 2G, 3G, 4G, 5G, 6G, and so on. Right now, we are fully deployed in 4G, with 5G gaining ground and 6G already under development. Wireless communications have enabled and increased the connection of billions of people to the Internet. Also, agreed standards for mobile phones allow people to use their devices everywhere. The development of mobile technology has opened a new market for business and personal uses.

THE INNOVATIVE LEAP – COMMUNICATION OVER THE AIR

In 1973, Dr. Martin Cooper, who was then the general manager at Motorola communications system division, made the first mobile phone call in public on a phone device that weighed 1.1 kg. Mobile phone calls happen like this: when we make a call, our voice is converted to an analog signal by the mobile's microphone. This analog signal is then converted to digital traffic and sent to the nearest mobile base station or called mobile tower through microwave spectrums.

The main feature of mobile technology is "portable." We also call mobile networks cellular networks because they are made up of "cells." We call areas of land cells that are typically hexagonal, with at least one transceiver cell tower within their area. The cells use various radio frequencies and connect to one another and telephone switches or exchanges.

There is a difference between cellular, like cell phones, and wireless technologies, like Wi-Fi. Cellular devices (smartphones, tablets, and mobile hotspots) come with a data plan and need cell phone towers to support

Internet access. On the other hand, Wi-Fi requires wireless devices (smartphones, tablets, and laptops) to connect to a router for Internet access.

NEW MINDSET FROM MOBILE TECHNOLOGIES

Technical Outlook – Starting from Spectrum

The radio spectrum allocated to mobile networks has expanded over time. Most mobile networks worldwide use portions of the spectrum given to the mobile service for transmitting and receiving signals. In each country, the particular frequency bands may also be shared with other radio-communication services, e.g., broadcasting and fixed wireless services.

For example, 4G LTE bands are discrete slabs of frequencies used for telecommunications. LTE Band 1 is stated to have a frequency of 2100 MHz (megahertz), but it uses frequencies between 1920 and 1980 MHz to uplink data and frequencies between 2110 and 2170 MHz to downlink data.

Another example is Citizens Broadband Radio Service (CBRS) which refers to 150 MHz of spectrum in the 3550–3700 MHz range (3.5–3.7 GHz). In 2015, the United States Federal Communications Commission (FCC) has designated for sharing CBRS among three tiers of users: incumbent users, priority access license (PAL) users, and general authorized access (GAA) users.

CBRS frequency band in the United States is significant because it enables different organizations (beyond telecommunication carriers) to make use of the 3.5–3.7 GHz radio spectrum to build their own wireless networks based on 4G LTE and 5G cellular technologies. Such a network is also called private LTE or 5G.

Business Outlook – Mobile Commerce

Examples of "wireless" have wildly different uses in business, such as radio and television broadcasting, radar communication, cellular communication, global position systems (GPS), Wi-Fi, Bluetooth, and RF identification (RFID).

Mobile technology's value is simply significant, as it has led us into a digital world in the true and large sense. It has made both life and business much easier. Mobile technology can save business time and money in many ways, for example, removing the need for expensive technology, such as landline network services.

Mobile commerce, also known as mCommerce, primarily uses wireless devices like cell phones and tablets to conduct commercial transactions online. mCommerce includes product trade and transactions, online banking, and paying bills. To operate mobile commerce, we can use advanced websites, dedicated apps, or even social media platforms like Facebook or Instagram that allow in-app purchases or linking to online stores. Mobile commerce volume is expected to hit $620.97 billion by 2024. This means nearly half (42.9%) of all ecommerce purchases will be made via a mobile device (source: https://www.shopify.com/blog/mobile-commerce).

Application Outlook – Internet of Everything

The Internet of Everything (IoE) extends the IoT emphasis on machine-to-machine (M2M) communications and describes a more complex ecosystem that also encompasses people and processes. IoE becomes possible because of the wide deployment and ubiquitous connectivity of communication networks, especially via wireless communications.

As one of the most important practices of future technologies, IoT is booming together with 5G. These include smart cities, the military, education, hospitals, homeland security systems, transportation, autonomous, connected cars, and agriculture. In cities, 5G will enable enhanced traffic management by supporting a massive number of IoT connections to traffic lights, cameras, and traffic sensors. Smart meters can be supported by 5G low-cost IoT sensors and connections and monitor energy usage and help reduce consumption.

While IoT is the network of physical devices where data collection and exchange occur without human intervention, IoE is the intelligent connection between people, processes, data, and things. IoE will lift IoT to the next level, and the benefits of using IoE include increased efficiencies, significant cost savings, and reduced wastage of energy. IoE enables people and their environment to be wirelessly connected by a continuum of AI-powered devices and networks, from driverless cars and search-and-rescue drones to implantable medical devices. Smart homes and cities today are pioneers of IoE.

INSIGHTFUL PRACTICE

The Incoming Wi-Fi 6

Wi-Fi as a type of wireless technology connects computers, tablets, smartphones, and other devices to the Internet and shares data online. Wi-Fi enables the radio signal to be sent from a wireless router to devices in range, bringing the devices online. Be aware that Wi-Fi is a wireless connection to a device, not the Internet itself. Wi-Fi serves as a linage extension for the Internet.

A Wi-Fi network is a simple and cost-effective way to connect with a Wi-Fi router or other Wi-Fi-ready devices wirelessly without wires. Wi-Fi lets you print pictures wirelessly or watch a video feed from Wi-Fi-connected cameras without needing physical connections. Wi-Fi is often used in cities or campuses for online access without a wired link. The Wi-Fi access points include devices like a wireless router or the private hotspot capability of another mobile device.

But keep in mind that the limitation of the Wi-Fi network is that it provides a connection in a limited area. Its radio connectively cannot reach beyond 20–25 meters. When your location is away from 25 m, you cannot get access to the Internet or local wireless network via WiFi.

The newest version of both Wi-Fi is Wi-Fi 6, the next-generation wireless standard faster than 802.11ac. Both Wi-Fi 5 and Wi-Fi 6 maximize the number of streams, thus enabling gigabit Wi-Fi speeds. Wi-Fi 6 can transmit 12 streams across the 2.4 and 5 GHz bands, compared to Wi-Fi 5, which has a limit of 8 in a dual-band configuration. More than speed, Wi-Fi 6 can provide better performance in congested areas, from stadiums to device-packed offices.

Wi-Fi 6 officially launched in late 2019, and its enabled hardware was released throughout 2020. Wi-Fi 6 has 160 MHz channel utilization capability and increases bandwidth to deliver more excellent performance with low latency. The WiFi 6 feature of Target wake time (TWT) which allows devices to sleep when there is no need for any router communication, significantly improves network efficiency and device battery life, including IoT devices.

Starlink Taking Over the Space

Starlink is a satellite Internet constellation operated by Elon Musk's firm SpaceX, providing satellite Internet access coverage to 31 countries on Earth to date with the ambition to extend its range to the whole world.

How does Starlink work? Starlink targets to launch thousands of mini communication satellites into low Earth orbit. From there, they will transmit fast Internet signals down to Earth. To date, 2,335 Starlinks have been launched, with 2,112 still in orbit. Once it becomes fully operational, Starlink can offer Internet access and services virtually anywhere on the planet.

Commercial-wise, instead of tiered plans, SpaceX offers a single Starlink Internet plan for everyone at USD$99 monthly with unlimited use. Such cost level places Starlink favorably at the affordable end of satellite internet pricing. How many Internet users can one SpaceX Starlink satellite support at once? Each satellite has about 20 Gbps capacity and can support 2,000 simultaneous customers at 100 Mbps.

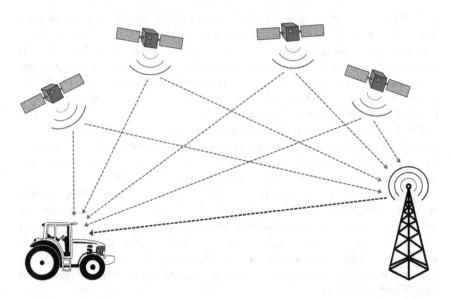

Satellites can support from the space many digital and wireless applications. e.g., GPS in our car.

What does a Starlink service kit look like? The hardware includes a satellite dish and router, which you will set up at home to receive the signal from space. Starlink can deliver high-speed broadband Internet service to locations where access has been unreliable or completely unavailable via traditional terrestrial infrastructure.

For comparison, fiber optical latency speed is around 17 milliseconds (ms) – slightly faster than copper cable Internet with 20–30 ms latency. The current latency for Starlink is about 20 ms and somewhat higher than fiber. However, Starlink latency is expected to be below 20 ms and reach under 10 ms.

Starlink orbits roughly 750 miles (1,200 km), much closer to Earth than GPS in medium-Earth orbit. This will help reduce the latency. Starlink satellite dishes can work fine on rainy, overcast days, but satellite signals can still be subject to heavy storms that affect the signal and slow the connection speed. Though it is even rarer, it is also possible for heavy storms to cause Internet access outages in the area.

Fixed Wireless 4G and 5G – The Hybrid Internet Service

Fixed wireless networking makes a fixed and mobile hybrid solution. Namely, it operates mobile devices in fixed sites such as homes and offices. Fixed wireless devices usually use their electrical power from utility mains, unlike mobile wireless devices, which are primarily battery powered.

Using wireless links between fixed points—a nearby tower and an antenna on a consumer's residence—4G and 5G fixed wireless service provides high bandwidth broadband connections wirelessly instead of through a wired cable connection. The Internet is sent through airwaves to the sites without requiring phone or cable lines (you can cut the cord). The service is reliable and fast – speeds are comparable to high-speed cable, with no data limits.

Fixed wireless is not a new technology, but its potential can now be maximized with 4G and 5G's arrival. 4G and 5G fixed wireless access aim to provide ultra-fast Internet/broadband services at ten times the lower latency rates and ten times the bandwidth across many radio frequencies in the future.

Fixed wireless 5G Internet is high quality. A good example is T-Mobile's Internet service via 5G to consumers in major cities. Unlike satellite Internet, it is low latency, and poor weather conditions do not disrupt service. It is a great option for rural telecommuters, businesses, and home Internet service.

5G is certainly suitable for fixed broadband in dense urban areas. But one thing to keep in mind, 5G has neither the capacity nor the cost-effectiveness to address the rural fixed broadband gap. 4G LTE is expected to continue serving rural areas in days to come.

Chapter 6

From Human Work toward AI – Bridge to the 4IR Future

Artificial intelligence is a software system that lets machine do human work by training the machines with big data via effective algorithms.

The notion of making robots has come along for quite a while, but the bottleneck is how to make really intelligent robots, and just for the robot to emulate a few human moves is not good enough. The answer is artificial intelligence (AI). We will talk further on AI in Chapter 17 when we cover machine learning (ML). The latest globally popular ChatGPT is an excellent example of the next generation of ML applications. ChatGPT's users can create any kind of text, including blog posts, essays, and program code, based on the generative AI models trained on vast amounts of information from the Internet, including websites, books, news articles, and more.

But ML is only the tip of the AI iceberg, which is the trademark of the 4th Industrial Revolution (refer to Chapter 2). AI indicates using computers to

DOI: 10.1201/9781003305187-7

do things that traditionally require human intelligence to handle. This means creating algorithms (refer to Chapter 16) to classify, analyze, and draw predictions from data. It also involves acting on data, learning from new data, and improving over time. Finally, AI needs robust computing chips and a cloud to happen.

AI is categorized broadly into three stages: ANI – artificial narrow intelligence, AGI – artificial general intelligence, ASI – and artificial super intelligence. The first or initial stage, ANI, is limited in scope, with intelligence restricted to only one functional area. ANI is like the infant stage of AI. AGI is at an advanced or adult level. This stage covers more than one field, like such as the power of reasoning, problem-solving, and abstract thinking. Some prediction says ChatGPT-5 is expected to reach AGI by the time of its release toward the end of 2023, which would be an interesting develoment worth following and watching. ASI is the final and highest stage of the artificial intelligence AI explosion, and in this stage, AI can will surpass human intelligence across all fields and become super AI.

Why is AI so important? In the 21st century, AI quickly evolves to become superior to humans in many tasks. However, does this imply we are ready to outsource our intelligence to technology? Some may feel AI will do our work so we can go lazy. This type of "slacker" approach is not the objective for AI. Do AI and human workers have the same qualities and abilities? In reality, they do not at least for the AGI (artificial general intelligence) stage. AI will augment human intelligence in most cases, not replace it. Once it gets to ASI (artifical super intelligence) level, however, the repercussions and impacts remain to be assessed.

AI technology is essential and pivotal because it enables and simulates human capabilities via software. Let software undertake the human intelligent work of understanding, reasoning, planning, communication, and perception increasingly, effectively, efficiently, and at low cost. A simple example is when you have some big data handy that is nearly impossible for human power to process, analyze, and make decisions quickly, hence AI, like deep learning, comes in to help. Deep learning makes collecting, analyzing, and interpreting large amounts of data faster and easier. It is highly beneficial to data scientists nowadays.

The transition from the ANI to the AGI stage has taken over a half-century. Right now, we are underway to complete the transition to the second stage – AGI, which aims that the intelligence of machines can equal humans. This is no doubt a significant achievement with unprecedented huge impact.

THE INNOVATIVE LEAP – DEEP LEARNING DRIVING AI

In the second AGI stage, we see a new generation of AI chips being developed for neural networks, deep learning, and computer vision. The AI hardware includes CPUs for scalable workloads, particular purpose built-in silicon for neural networks, neuromorphic chips, etc.

What leads the trend is the maturing of ML, supported partly by cloud computing resources and widespread, web-based data gathering. As we speak, ML is propelled forward by "deep learning," which is via a method called backpropagation.

ML targets computers' ability to think and act with less human intervention and assistance. In contrast, deep learning is taking a further step about computers learning to think using structures modeled on the human brain.

As mentioned above, one of the famous deep learning algorithms is backpropagation, short for "backward propagation of errors," for supervised learning and training artificial neural networks. It is the practice of fine-tuning in assigning neural net weights based on the error rate (i.e., loss) from the previous iteration. Proper tuning of the weights ensures lower error rates, making the model reliable by increasing its generalization.

AI is impacting our life and changing virtually every industry and area. Emerging technologies such as big data, robotics, and IoT are also boosting AI, which will continue to act as the innovation driver in this digital era.

NEW MINDSET FROM AI

Technical Outlook – Deep Learning

Deep learning is an advanced AI technology where data engineers train machines by feeding them different kinds of data. As time goes by, the machine learns to make decisions, solve problems, and perform other types of tasks on its own based on the data set given to them.

Deep learning networks learn by discovering intricate structures and patterns in the data they experience and then interpret the data and build computational models with multiple processing layers and levels of abstraction. Deep learning is trying to replicate the human brain working process, namely, transformation and extracting various features. The models attempt to establish a relationship between stimuli and associated neural responses in the brain.

Therefore, it requires considerable data to perform better than another human brain. Deep learning is costly to train due to complex data models. Moreover, from the hardware angle, deep learning requires expensive GPUs and hundreds of machines. This would increase the cost to the users.

In scenarios where there is a lack of domain understanding for fixed features and patterns, deep learning outshines others, which means deep learning has advantages in complex problem solving such as image classification, natural language processing, and speech recognition.

Business Lookout – Adopting Generative AI from Deep Learning

Generative AI is a type of deep learning called generative adversarial networks and has a wide range of applications, including creating images, text and audio. It can generate new images, videos, or text based on training data. The latest ChatGPT is a great success case of generative AI.

From creating new and original content to revolutionizing industrial practice, generative AI has the potential to shape the future in countless ways, such as creating new forms of art and expression, improving healthcare outcomes, making investment decisions, and so on.

From the angle of algorithms, generative AI means generating new outputs based on the data they have been trained on. Unlike traditional AI systems that are designed to recognize patterns and make analytics or predictions, generative AI creates new content with new values, which makes generative AI a more powerful tool in our life and work.

By adopting and deploying the right AI technology, such as generative AI, businesses may save costs by automating and optimizing routine processes and tasks, lifting productivity and operational efficiencies, and accelerating business decisions based on AI outputs.

Application Outlook – AI Power in Practice

AI can help solve complex issues in various industries, such as entertainment, education, health, commerce, transport, and utilities.

We can group AI applications into five categories:

Reasoning	Solve problems through logical deduction. For problems such as financial asset management and application processing, legal assessment, weapons systems automation, online game playing
Knowledge	Present knowledge and report about the market, such as financial market trading, fraud prevention, drug testing, medical diagnosis, media polls
Planning	Establish and achieve goals, such as inventory management, demand forecasting, digital network optimization, transport navigation, delivery scheduling, logistic arrangement
Communication	Understand spoken and written languages, such as translation of foreign languages, real-time transcription, intelligent assistants, voice and data integration
Perception	Infer things about the objects or targets via sounds, images, and other sensory inputs, such as medical diagnosis, autonomous vehicles, surveillance

Automated driving makes a famous AI deep learning use case nowadays: it helps researchers and testing drivers automatically detect road objects, such as stop signs and traffic lights. In addition, deep learning can do an excellent job of detecting pedestrians, which helps decrease accidents.

INSIGHTFUL PRACTICE

Google's AlphaGo and AI

AlphaGo is an AI player specializing in Go, a Chinese strategy board game. Its rivals are human competitors. AlphaGo represents a Google DeepMind project that takes the ability to beat a human player at strategic games as a measure of AI development. That makes AlphaGo special because while it was designed, named, and trained to play Go against the masters, its potential functionalities go well beyond the realm of board games, unlike most of its AI contemporaries.

The goal of AlphaGo is an attempt by the Google Deepmind project to design and engineer a system, a new set of algorithms and techniques. The objective is to challenge the professional human Go player and even the world's top Go player champion. The project did reach human-level intelligence in beating all five games of Go.

AlphaGo had a record of beating European champion Mr. Fan Hui and became the first program to defeat a professional player. There was only once for Lee Se-dol, a South Korean Go player, to date, ever beat the AlphaGo. In 2016, Lee, during a five-match showdown against AlphaGo, lost four times but beat the computer once.

AlphaGo and its new versions use a "Monte Carlo tree search algorithm" to figure out best moves based on knowledge previously collected by ML, specifically by an artificial neural network which is a deep learning method using by extensive training, both from human and computer play.

AI will Fight the Next War

From the military standpoint, AI technology is central to providing robotic assistance on the battlefield, enabling forces to maintain or expand war-fighting capacity without increasing the workforce. Compared with conventional systems, military systems equipped with AI can handle larger volumes of data more efficiently. Additionally, AI improves combat systems' self-control, self-regulation, and self-actuation due to its inherent computing and decision-making capabilities.

While the US military is planning to leverage the computer vision capability of AI to hunt down submarines, detect an enemy intrusion, or decode messages using machine learning ML abilities, several countries around the world have given the nod to using AI-powered drones in the name of national security.

The US Airforce has experimented with using deep-learning AI as flight simulators, letting the algorithms show if they can match the skills of veteran human pilots in grueling dogfights. The Airforce says AI pilots will only be used as "wingmen" to real humans when they are ready to be deployed.

AI's incarnation motivates global powers to prepare themselves to control and maneuver advanced technologies. Army soldiers can also use an AI engine which provides information on the surrounding environment and terrain. With efficient information regarding the environment and terrain, Army can plan its appropriate activity and outputs.

So, AI will be a super soldier for the future of warfare. It will increasingly process defense-related information, filter such data for the greatest threats, make defense decisions based on its programmed algorithms, and perhaps even direct combat robots.

AI and Smart Home

A smart home equips a convenient home with digital appliances and devices. Users can control the devices remotely from anywhere with an online connection via a mobile or other networked device. Smart Living is a trend with smart home capabilities that enable people to enjoy the benefit of new ways of living. Smart homes aim to make life more efficient, controllable, economical, productive, integrated, and sustainable.

Smart living with AI and integrated IoT tools is no longer a luxury as home automation setup has become more affordable. Besides its role in home security systems, AI can also control smart devices with voice AI-enabled agents, such as Alexa, Siri, and Google Assistant. These devices own a higher intelligence that assists people to better manage their daily tasks at home from anywhere in the world.

For example, smart kitchen appliances with good recipes can help automate crockpots and rice cookers for delicious meal cooking; People can control the process with Alexa and other connected devices. Using a smart kitchen, the users could cook the food they choose or heat it before getting home from outside.

We can look at the top five features of a smart home system via AI.

- Advanced home network. This refers to the "Internet of Things" for now.
- Remote access. People can obtain remote access to one's home systems.
- Advanced security.
- Lighting control, automated window treatments.
- Distributed audio

Overall, smart home devices have made a highly positive impact on the majority of our lives. With entertainment and safety devices being most popular among most of us, it is no surprise that smart home devices have made us happier at home and increased our feeling of security.

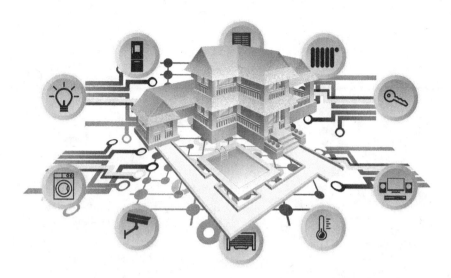

Smart Home is to use up-to-date digital technologies, IoT, sensors, and Wi-Fi to automate home control in terms of security, temperature, cooking, and other tasks.

Layer 2 – Digital Thinking from the Building Pillars of Technologies

How do you cultivate digital ways of thinking? Layer 2 of the Cognitive Model of Digital Transformation suggests starting your digital thinking from the pillars of digital technologies. These pillars are the core and critical solutions that forge the digital technology ecosystem. Once you understand how cloud computing, fiber optics, the world wide web, virtualization, artificial intelligence, and cybersecurity work, you will start to think things along the digital ways and expect how they would disrupt and replace the traditional ways of thinking and doing things.

> Like air and drinking water, being digital will be noticed only by its absence, not its presence.
>
> – Nicholas Negroponte, Author of the Best-seller *Being Digital*
> (Alfred A. Knopf, 1995)

> We've moved from digital products and infrastructure to digital distribution and Web strategy to now into more holistic transformations that clearly are based on mobile, social media, digitization, and the power of analytics, and we think it's really a new era requiring new strategies.
>
> – Saul Berman, Chief Strategist, VP & Interactive
> Experience Partner of IBM

DOI: 10.1201/9781003305187-8

Layer 2 – Digital Thinking from the Building Pillars of Technologies

Chapter 7

From Electronic Tubes to Microchips – Brain Cells of Digitalization

Microchips are electrics built on silicon material and operate as the operate as the brain cells of modern digital applications, devices and equipment.

Why do we have an area in Northern California, called Silicon Valley? Silicon is from minerals that make up 90% of the Earth's crust. A microchip (a chip, a computer chip, an integrated circuit, or IC) is made of electronic circuits printed on a small flat piece of silicon made from silica sand. Microchips are the brain cells of information and digital technologies.

Micro means the chips are the incredibly complex yet tiny modules that store computer memory or provide logic circuitry for microprocessors. Most older people might have opened a radio box and seen what was inside: many electronics powered by vacuum tubes. But that era was long gone, and microchip technology changed the world. Electronic devices received a historic boost with the invention of the silicon transistor, marking the beginning of the end for bulky electronics via vacuum tubes.

DOI: 10.1201/9781003305187-9

The pioneers known for inventing microchip technology are Jack Kilby and Robert Noyce from the USA. In 1959, Kilby of Texas Instruments succeeded in miniaturizing electronic circuits, and Noyce of Fairchild Semiconductor Corporation also invented a silicon-based integrated circuit (IC). This invention would revolutionize the electronics industry, helping make cell phones and computers possible and widespread today.

In the digital era, a small electronic chip will drive the world. Perhaps the best-known chips are the Pentium microprocessors from Intel used in our PCs. Microchips exist in almost every electronic device we use today. Why is a smartphone innovative? Because it is an SoC, or system-on-a-chip, which is the smartphone's brain. Combining multiple components into a single chip save on space, cost, and power consumption. Essentially, a SoC handles everything from the Android operating system to detecting when we press the power off button.

Another great example is that Tesla uses several chips inside its vehicles for different control systems and its infotainment system. Most famously, the automaker uses a chip designed for the well-known self-driving software. So, who is doing the auto-driving for Tesla? The chip.

THE INNOVATIVE LEAP – GETTING MICRO IN CHIPS

The microchip has miniaturized computers, communications devices, controllers, and hundreds of other devices. Since 1971, whole computer CPUs (central processing units) have been placed on microchips. In contrast, such a CPU in the past was a mainframe that could fill an entire room.

Technically, a computer chip consists of a series of electronic circuits installed onto a conducting material, usually silicon. The components etched onto a microchip include transistors, capacitors, and resistors. A transistor is an essential device capable of amplifying and switching electrical signals. A capacitor temporarily stores electrical charges, while a resistor controls current by providing resistance.

Each chip contains many transistors that make up a processor. Thus, one chips can take tens of millions of transistors that are aligned together to generate an electrical signal. Several chips can be assembled with different amounts of memory storage space in a central processing unit. Microchips forge the building blocks to make computers and run the software in the size of a nanometer (nm). As a reference, one nanometer is only one billionth of a meter, as tiny as the wavelength of visible light. The diameter of an atom routinely ranges from about 0.1 to 0.5 nanometers.

In 2012, Intel launched its 22-nm processor, which then was the world's smallest and most advanced transistors in production. In 2014, Intel developed an even smaller, more powerful 14nm chip; today, the company is working to bring its 10-nm chip to market. In May 2021, IBM announced it had created a 2-nanometer chip, the smallest, most powerful microchip yet developed. Most computer chips powering devices today use 10-nanometer or 7-nanometer process technology, with some manufacturers producing 5-nanometer chips.

NEW MINDSET FROM MICROCHIPS

Technical Outlook – Small Moves Big

In 1965, Mr. Gordon E. Moore, the co-founder of Intel, initiated the famous Moore's Law. It indicates Moore's observation that every two years, the transistors on a microchip double in number, and the cost of computers is halved. Moore's Law predicts that the speed and capability of computers we use will increase every couple of years. Meanwhile, we will pay less for them.

If electronics now use half the time to make a calculation, that means the chip is twice as fast. But the shrinking cannot go on forever, and we are already starting to see three interrelated forces—size, heat, and power—threatening to debate or deviate from Moore's Law.

Breakthroughs will be in need. Quantum computing can be the answer. Compared to the semiconductor integrated circuit as the microchips we use today, quantum chips are a new generation of technology. Quantum computing will be 158 million times faster than the most advanced super-computer we have now. In other words, quantum computing can do what it would take traditional supercomputer years to accomplish in a few seconds.

Business Outlook – Moore's Law Effects

We face challenges and benefits from the vision of an endlessly empowered and interconnected future. Minimizing microchips and transistors, as the new technology approach we are adopting, has propelled advances in computing for more than half a century.

While we must cope with the growing concerns around privacy and security regarding digital technologies, the advantages of ever-smarter computing technology can benefit us in the long run. It helps keep us healthier, safer, and more productive. Instead of physical processes and hardware, applications and software will help improve the performance of computers. Computer innovations are driven by advances in cloud computing, 5G communication, the Internet of Things (IoT), and quantum physics.

As Moor's law predicts, as transistors in ICs become more efficient, we expect computers to become smaller and faster, meanwhile its cost has been decreasing annually, partly because of lower labor and semiconductor costs from mass productions.

Application Outlook – Computerizing All Sectors

Microchips are made for program logic (or microprocessor chips) and computer memory (or RAM chips) and are used for special purposes such as analog-to-digital conversion, bit slicing, and gateways.

In practice, every facet of a high-tech society enjoys the benefits of massive computing. Mobile devices, including smartphones and computer tablets, become possible due to tiny processors; By the same token, video games, spreadsheets, accurate weather forecasts, and global positioning systems (GPS) also run over microchips.

As time goes, smaller and faster computers improve transportation, health care, education, and energy production for more use cases. Nearly all industries have progressed because of the increased power of computer chips.

INSIGHTFUL PRACTICE

x86, GPU, and Intel

Intel, or the Intel Corporation, is an American manufacturer of semiconductor computer circuits. It is headquartered in Santa Clara, California. The company's name comes from "integrated electronics." Intel is one of the chip-making giants, alongside Taiwan Semiconductor Manufacturing Co. and South Korea's Samsung Electronics Co.

Intel is the world's largest semiconductor chip manufacturer by revenue and is the developer of the famous x86 series of microchips found in most personal computers (PCs). Intel manufactures graphics, flash memory, motherboard, and other computing devices.

The famous x86 is a generic name for the Intel microprocessor families used for PCs since 1982. It got the name x86 because it began with the 80286 microprocessors. x86 microprocessors can run almost any computer, from laptops, servers, desktops, and notebooks to supercomputers. Today, the term x86 refers to any 32-bit processor compatible with the x86 set.

GPU (graphic processing unit or graphic card) is another key Intel product set. It is a single-chip processor used chiefly to manage and enhance video and graphics performance. GPU is critical when we play video games or watch online streaming or videos like Youtube. Processor numbers with a "G" are optimized for graphics-based usages and include newer graphics technology. The suffix indicates the level of graphics offered by the processor; higher numbers (e.g., G7) indicate improved graphics performance relative to lower numbers (e.g., G1).

Intel has a strong competitor who is another major microchip maker called AMD, which is short for Advanced Micro Devices; like Intel, it produces more than just microprocessors. Both companies create motherboards, servers, and other computer-related hardware.

Intel has been providing x86 series of microchips for our PCs since early 1990s.

Ultra-Small – Nanochips

Nanotechnology is to produce new structures, materials, and devices on a near-atomic scale. How small can it be? Nanomaterials can have a length scale between 1 and 100 nanometers. Nanotechnology includes engineered structures, devices, and systems.

In 2000, the semiconductor industry invented nanochips with features measuring less than 100 nanometers, and these devices are found in average desktop computers nowadays. A microchip is different from a nanochip in that it requires a lot of power, while a nanochip is much smaller and requires little power. The two chipsets are similar in size, but their manufacturing process differs. The nanochip is smaller than a microchip, making it much easier to manufacture. It also requires far less energy and can operate at a higher speed.

A nanochip is so small that only individual particles of matter can play significant roles. Smaller electronic and computer components have always been the industry's primary goal. Today, engineers can place such a computer component inside a microscopic capsule. The chips, called motes, are the size of dust mites, measuring less than 0.1 cubic millimeters, and they can only be seen under a microscope. Motes operate as a single-chip system, complete with their electronic circuit.

Healthcare is a major arena where nanotechnology is used. The applications of nanomaterials as components or as additives for various medical devices have led to improved biocompatibility as nanorobots, novel nanochips, and nano-implants serving, for example, cell repair. In the future, nanochips may even be used in surgical procedures, such as heart valve replacement.

What Enables a GPS Tracker?

GPS-powered navigation or maps are now widely used by regular drivers worldwide. A GPS tracking unit, geo-tracking unit, or simply tracker is a navigation device, generally on a vehicle, asset, person, or animal, that uses the Global Positioning System (GPS) to determine its movement and geographic position (geo-tracking) to determine its location.

A Global Positioning System (GPS) embeds a microchip, which receives signals from four or more of the 27 GPS satellites currently orbiting Earth. A spot positioning needs four satellites because each data from one satellite puts the spot in a sphere around the satellite. The system can narrow the possibilities by computing the intersections to a single point. Three satellites' intersection places the spot on two possible points. The last satellite settles the exact location. A chip in the user's phone, or any GPS device, can receive those signals at least three times. Then it can pinpoint where the user is. Current civilian GPS receivers are accurate to 10 to 33 feet, depending on conditions.

Once the micro GPS chip receives GPS signals from satellites via wireless connectivity, it transmits the data to a terminal device like your PC or mobile phone. Regular users of GPS include scientists, pilots, boat captains, surveyors, first responders, and workers in mining and agriculture. They use GPS information to prepare accurate surveys and maps, take precise time measurements, track position or location, and navigate.

Chapter 8

From Dumb Pipe to Next-Gen Intelligent Network – Auto-Healing Infrastructure

The network for Internet is global powered by software, virtualization, LTE and 5G, and fiber optics.

How important is a network for digitalization? Well, no network, no digitalization, period. The network is the bloodline or backbone of digitalization. Simple examples include, for a smartphone to receive or send data, we need a network to traverse the traffic via wireline or wireless network.

Talking about network nowadays may entail many things: copper vs. fiber, narrowband vs. broadband, wireline vs. wireless, 4G vs. 5G, dumb pipes vs. intelligent networking, etc.

Copper vs. fiber is about the material we use for the network. Fiber optic transmits data via a light or at the speed of light, while copper delivers information as it carries electricity. Copper-based transmissions currently max out at 40 Gigabits per second. In contrast, the bandwidth limits

DOI: 10.1201/9781003305187-10

imposed on fiber are typically none and, based on tests, can be in hundreds of terabits per second. Fiber is the future.

The key to fiber optical transmission (refer to Chapter 11) is converting electrical signals to optical signals. For example, you sit at home, using Verizon FiOS for high-speed internet access. FiOS is known as an FTTP (Fiber to the premise) technology. All the links are fiber until once into your house; the last few feet to your router remain copper lines already wired at your home. The signal will be converted between electric and optical via a small onsite box (called a NID – network interface device) to make the copper line work with fiber.

For narrowband vs. broadband, the US Federal Communication Committee (FCC) has regulated that a broadband Internet connection needs to be 25 megabits per second or above. Only some copper-based old technologies like ADSL or ISDN may go below 25 Mbps in service, namely, narrowband. Any fiber-based Internet access should automatically qualify for broadband services. Why do we need broadband Internet access? Well, the explosion of Internet applications is the major driver.

As for wireless vs. wireline, there is a myth to decode here. People tend to think wireless means end-to-end, with no wireline involved. For example, Mr. Adam in New York City calls Ms. Eve in Washington DC via a smart cell phone. Is this communication process totally wireless? Accurately speaking, only the last mile (from the cell phone to the mobile tower nearby) is wireless. The remaining part is still wireline, mostly over fiber core networks nowadays.

Why is the wireless architecture designed like this? Because that is the most quality, time, and cost-effective way to handle communications. Then some folks may ask, what about 5G? Is 5G still under a similar structure? The answer is yes, and actually, 5G would consume more fiber networks than the previous 3G or 4G.

THE INNOVATIVE LEAP – NEXT-GENERATION INTELLIGENT NETWORK

In essence, the next-gen network will become intelligent and self-healing, capable of four Cs – collecting data, cognizing information, controlling the traffic, and collaborating with different systems.

The key here is the network being intelligent and convergent. In terms of services, an intelligent network can distinguish and support specific but different applications converged over the same core network. Network management means automation in troubleshooting or healing, maintenance and upgrade, billing, and marketing data analysis.

What three pillars make an intelligent network possible? SDN, NFV, and AI/ML are all under the software-defined architecture (SDA). SDN stands for a software-defined network that can do centralized control and management of a global or national network.

NFV means network function virtualization that can generate many network functions in the form of software and virtual fashion, making network control much easier and more effective. Finally, AI – artificial intelligence and ML – machine learning are all software algorithm driven and highly cognitive solutions that can be used for network services and management.

This leads to a cognitive network that uses cognitive processes to detect and understand current internal conditions. It makes decisions based on its findings and then learns from those decisions. Cognition is a psychological term and "the action or process of acquiring knowledge, by reasoning, intuition, or the senses. "senses."

By making the network cognitive, we allow the network itself to observe and act. The network can find resource usage patterns and take actions to optimize this usage. Thus, a cognitive network achieves a higher level of automation where the human network operator is relieved from network management and configuration tasks.

So, the modern intelligent network is armed to the teeth with cutting-edge innovations and new technologies. Software is replacing hardware, intelligence is replacing dummy management, and super high speed pipe is replacing slow circuits.

NEW MINDSET FROM NEXT-GEN INTELLIGENT NETWORKING

Technical Outlook – New Network Architecture

How do we define a next-generation network (NGN)? It is an Internet Protocol (IP) packet-based network that provides telecommunication services to users and uses multiple broadbands, QoS (quality of services) – enabled transport technologies. NGNs commonly operate around the IP layer; therefore, it's also called "all IP." NGN makes unfettered access possible for users to networks and to compete for service providers and services of their choice.

NGN's service-related functions are on the IP layer independent of the underlying transport-related technologies. Hence NGN is also called overlay network architecture, or say it is an SDN overlay adding in network virtualization. SDN involves running a logically separate network or network component on top of existing infrastructure.

Such overlay networks have several advantages, enabling developers to easily create and implement protocols on the web, from data routing to file sharing management. Additionally, data routing in overlay networks provides great flexibility with support for multi-path forwarding.

Business Outlook – Consolidation and Convergence

As described above, today telecommunication networks develop toward All-IP NGNs. This global migration is driven by higher economies of scale and scope, cost savings, opportunities for new products, and an improved experience for the end users.

One network provider can converge and deliver voice, data, and video networking services in a single network fabric, in contrast to providing a separate network for each of these services in the past. This allows an organization to use one network from one single provider for all communication and cloud-based services.

The NGN will also converge seamlessly with 5G, enabling network slicing. This allows the network vendors and operators to split the 5G network spectrum and serve various use cases, including mobile networking, smart homes, IoT, and smart entry grids. In other words, 5G supports numerous diverse use cases that the same physical NGN will serve.

Application Outlook – Doing More with Less

Next-Gen intelligent networking can provide all-in-one solutions, including VoIP, presence-based applications, instant messaging, IPTV, and location-based services. These services are provisioned on the public Internet or private IP networks, and access is available from underlays, including traditional circuit-switched networks.

Significant speed improvements in both wireline and wireless network technology are underway. This will bring about new network use cases and applications. For instance, improvements offered by 5G will give businesses access to lightning-fast data transfer speeds and improved network reliability.

Network capacity means the maximum amount of data reliably transferred between different locations over a network. Higher bandwidth means businesses can achieve a higher data transfer rate, leading to shorter download times. This is especially significant for companies that often need to transfer and download large files.

Another goal of the next-gen network is to reduce the operational complexity with automation or called self-healing. 5G, for instance, brings significant benefits to enterprises and service providers, but it also increases complexity that impacts speed, efficiency, and profitability. Network automation is essential to manage such complexity. You can significantly reduce manual workloads and eliminate human error with network automation.

INSIGHTFUL PRACTICE

Self-Healing Network – A Dream Coming True

Network automation (self-healing) uses software to automate network and security provisioning and management. Network automation is geared to reduce human errors and continuously manage the network by itself to maximize network efficiency and functionality.

Telecom service providers embrace network automation in which software automatically configures, manages, and tests network devices. Hence the three major benefits of network automation include business agility, operations speed, and OPEX reduction.

Human error can cause significant network downtime and outages. Such errors have caused about 75% of data center failures. Now an automated network can boost reliability, performance, and security. Network automation uses programmable software to manage network resources and services, and it allows network operations (NetOps) teams to configure, scale, protect, and integrate network infrastructure and application services more quickly than when performed manually by users. Thus, it allows the speedy rollout of new services, devices, and applications based on established and reusable processes.

Network automation can help scale a client's IT infrastructure, even as it runs complex analyses based on inputs from the various devices available within the client's network. By using automation programs, the operator will have a perfect network view and can easily access network data and performance reports, improving the network's control. With the vision of network automation, the operator can also develop new products and services.

A Next-Gen NOC for Better Customer Experiences

As Telecoms look to migrate their networks to 5G to drive new services, they must manage customer expectations for both the latest and the existing services. The cost of deploying and the complexity of managing these new services while maintaining an existing network are also challenging.

To meet this challenge, it all starts with examining the Network Operations Center (NOC) and how it operates today, looking at how it can be transformed for the future. A NOC is where enterprise ICT administrators – in-house or third-party outsourcing – supervise, monitor, and maintain a telecommunications network.

A next-gen NOC offers a modern, modular, highly agile, and scalable framework for enterprises to expedite the business needs of today's cost-conscious, agile, and competitive telecom market. It is powered by cognitive intelligence and automation that benefits the delivery of improved service and revenue growth. Such a NOC is driven by performance optimization and lower operational costs.

Those professionals working in a NOC are called NOC Specialists, or network operating center specialists, who installs, manages, and tests the systems that connect computers in an organization. Many NOC specialists work in the telecom and IT industries, helping firms solve a range of technical problems while also providing B2B customer support.

As network automation progresses, we still need centralized NOCs to coordinate the network management, just for smaller sized and smarter NOCs to handle more work and cover larger global scale network.

SD-WAN – A Breakthrough in Next-Gen Network

A wide area network (WAN) is not tied to a single local location. Rather, WANs link separate LANs and enable communication, information sharing, and more services between devices worldwide. Through a WAN service provider, WAN plays a major role in the art of the communications network.

An SD-WAN is a virtual WAN architecture that allows enterprises to use any combination of underlay transport services – including MPLS, LTE, and broadband Internet services – to connect users to applications securely. SD-WAN separates client's applications from the underlying network with a policy-based, virtual overlay.

SD-WAN eliminates those waiting lags, enabling the client to set up a new broadband connection in days. In the past, ordering 100 Mbps Ethernet circuits might take months to install. Speed of deployment is undoubtedly an essential benefit as companies transition much of their workforce to remote locations.

The key strength of SD-WAN is that it uses a centralized control to securely and intelligently route traffic across the WAN to cloud-based SaaS and IaaS vendors. This capability increases application performance and delivers a high-quality user experience. Consequently, SD-WAN can increase business productivity and agility and reduce IT costs.

As the next-gen network solution, SD-WAN automates many on-premises and cloud network tasks that used to be typically done by humans. SD-WAN is bringing about an intelligent, self-learning future network that can make traffic routing decisions while balancing various workloads in real-time.

SD-WAN leading the next-gen network trend.

Chapter 9

From On-Prem IT to Cloud Computing – Powerhouse for ICT

Cloud computing is remote computing resources over virtual machines and distributed over public or private networks.

Cloud computing has been a buzzword since the mid of the first decade of the 21st century. It has revamped the landscape of information and communication technologies (ICT). Today, you can still get somehow digital without cloud computing. But cloud computing is a must for digitalization – entailing the large scope or hyperscale of digital applications in service productions, business operations, and daily lives.

What is cloud computing? Conceptually, cloud computing means the centralized services of IT infrastructure, resources, and applications or running IT resources and operations remotely. A good analogy is the wide

DOI: 10.1201/9781003305187-11

utilization of electricity from the late 19th century to today's comprehensive functions and benefits of cloud computing.

In 1882, Thomas Edison started to form the Edison Electric Illuminating Company of New York, which brought electric light to parts of Manhattan. By 1925, most Americans bid farewell to gas lights and candles at home, and half of all homes in the United States had electric power. But keep in mind that light at home is only one of the applications of electricity.

The key here is more about how electricity is produced and distributed. Electricity can be produced via coal, water, oil, and nuclear. When electricity started for industry use, each factory would operate an electric generator on site by themselves, meaning the power source was produced, controlled, and managed locally and onsite in a fragmented way by each individual business organization.

Then a new notion came up: since every factory or plant used electricity, why not set up a centralized electric plant specializing in power production, transport, and supply? In this way, all the individual factories could plug into the electric outlet over the power lines from the central power plant and use and pay for the electricity as they need and as a service. This could lift productivity for all parties exponentially: the power plant specializes in power production, and other businesses focus on what they do in their specialty. This so-called "utility model" remains the concept of how we use electricity to date.

THE INNOVATIVE LEAP – THE VIRTUAL AND LIMITLESS COMPUTING

By the same token, as electricity is the energy power of industrialization, cloud computing is the energy plant for IT and digitalization. We can make computing resources as a utility as well. Here by "energy plant," we mean the robust computing, storage, networking, and security capabilities housed and produced by cloud computing.

Organizations in the past often ran their IT operations (including servers, systems, switches, storage, applications, security, etc.) all on-premise or from their own data centers, just as factories used to keep a power generator onsite. But now, cloud computing has started to change the landscape and serve as a centralized power factory of ICT and digitalization in a more cost, time, and resource-effective way.

However, there is one major difference between using electricity and cloud computing. For electricity, all applications are at the edge or user end. For instance, in our homes, we use light bulbs, TV, Internet, refrigerators, PCs, washing machines, electric stoves, microwave ovens, air conditioners, etc., all powered by electricity. Only we do not produce electricity at home.

For cloud computing, on the other hand, not only are basic IT resources and infrastructure residing in the cloud, but applications can also reside in or near the cloud. While a power line only distributes power to our homes, the IT and telecommunication network also distribute applications to our offices and homes from the cloud.

For example, if I sit here using Microsoft Word, I have two options: the first is to install the Microsoft Office software on my PC and use Word from there. But this is becoming the old fashion. I can also use Word by subscribing to Office 365 online and using it from the cloud. This is the new way to go.

As long as we have a reliable, convenient, and robust network (refer to Chapter 8), using Office 365 online from the cloud would provide us with many more benefits than we use Word from our PC. Office 365 brings about more scalability and agility: we can use and save any large files to Office 365 without worrying about our local PC disc sizes. Also, now we do not have to load our PC with many software versions and can save a lot of processing and memory power for the applications. Office 365 also offers and orchestrates many new features like MS Teams we can use.

For up-to-date software management, Microsoft can now upgrade and patch Office 365 as needed anytime and always keep the capabilities up-to-date, unlike before we had to purchase new version Word discs to upgrade the software; it is also more cost-effective: now we do not need each time to buy an Office disc for upgrading. Instead, we subscribe to Office 365 and pay it per use or on demand. This approach also represents the benefits we make use of cloud computing in general.

NEW MINDSET FROM CLOUD COMPUTING

Technical Outlook – Computing as a Utility

In the cloud era, customers can self-provision computing resources without the 3rd party intervention, i.e., by the cloud service provider (CSP). The client's demands, such as server time, storage, network, and application, are all accessible from the cloud. CSPs typically pool the computing resources such as storage, processing, and network. This way, the CSPs can serve multiple customers or via the so called multitenant model. CSPs can dynamically assign computing resources to different customers on demand.

When using cloud computing, customers may need to learn the physical location of the computing resources. At a higher level of abstraction, the location of the cloud resources can be specified such as a country, state, or data center. More imporatantly to the users, the computing resources are accessed over the network by using different client platforms such as mobile phone, laptop, desktop, and tablet.

Some cloud customers may think the computing resources from the cloud appear to be unlimited. It is not infinite, but cloud resources can be rapidly and elastically provisioned to meet customers' demands. It is called cloud scalability. Customers can demand and purchase computing resources anytime and in any quantity.

Business Outlook – Cloud First

Many organizations have started a cloud-first strategy that allows businesses to save capital expense (CAPEX) and operation expense (OPEX) on software, platforms, and infrastructure. The clients will not need to build their own cloud stack. Instead, they subscribe to a CSP that can provide business-class cloud services at a cheaper cost.

Cloud computing allows storing and accessing data and programs over the "cloud" remotely instead of on your computer's hard drive or on-premise IT operations. This leads to flexible resources, faster innovations, and economies of scale. The future of cloud computing can be a hybrid IT solution – a combination of cloud-based software solutions and on-premises computing. The hybrid cloud is scalable and flexible, backing up security and control from the data centers.

Edge computing is another emerging cloud trend. It will store and process data and information in localized data centers closer to the devices/apps that use them. Edge computing means computing and management to be handled locally rather than at the central cloud network. Edge computing is especially useful in remote locations for big data and workload processing with little connectivity to a remote and central cloud.

Application Outlook – IaaS, Multi-Cloud, and Metered Usages

Infrastructure as a Service (IaaS) can make a strategic move for businesses to adopt cloud computing. IaaS delivers on-demand essential resources to enterprises, including computing, network, and storage. Such a utility model – using a central infrastructure over the Internet on a pay-per-use scheme – is an obvious choice for companies. It helps save on acquiring, managing, and maintaining an IT infrastructure. Enterprises can keep some mission-critical tasks on-prem and migrate others to the cloud, making a more effective hybrid model.

Multi-cloud allows businesses to use multiple clouds from different cloud providers for diversity or backup. The approach can make any mix of Infrastructure, Platform, or Software as a Service, namely, IaaS, PaaS, or SaaS. Multi-cloud allows businesses to choose which workload is best suited to which cloud environment and also helps to avoid vendor lock-in.

Cloud computing also provides metered on demand service: the computing resources used by the customers can be metered at some level of abstraction, depending on the type of service. The utilized resources can be reported with metering capability, providing transparency between the provider and the customer. This becomes exactly how the utility model bills the customer – you only pay for what you use.

INSIGHTFUL PRACTICE

SaaS – Software from the Cloud

Software as a Service (SaaS) uses a subscription model and delivers software services over the network and as a service.

SaaS is a cloud-based software distribution model that brings agility and cost-effectiveness to companies. It is also attractive to businesses due to its simplicity, easy user accessibility, enhanced security, and widespread connectivity.

SaaS gives service access to cloud-based apps over the Internet instead of the traditional local disc installation. SaaS is all remote and virtual through the cloud delivery model. Bidding farewell to install and maintain software, users can access it via the Internet, freeing themselves from complex software and hardware management.

SaaS provides a new and complete software experience. As briefly mentioned previously, Office 365 is a successful SaaS, and it provides MS Office Suite (Office Web Apps) online along with Windows 10 Enterprise, SharePoint Server, Exchange Server (email), and Lync Server (for unified communications). The hosting cloud platform used by Office 365 is MS Azure, which also makes the Windows Server operating system and many other features available as services.

Such SaaS cloud-based service offers a monthly or annual subscription to access all the Microsoft tools and apps hosted on Microsoft servers called MS OneDrive. Users can access MS OneDrive from virtually anywhere through the Internet, and the package offers additional features. OneDrive can also be purchased as a standalone service.

MS Azure and LinkedIn

MS Azure, as the public cloud patfrom offered by Microsoft provides a range of cloud services, including computing, analytics, storage, and networking. The Azure cloud platform owns more than 200 products and cloud services designed to help enterprises bring new solutions to life. The Azure cloud also allows to operate applications across multiple clouds, on-premises, and at the edge. You choose the tools and frameworks you like.

For enterprises with MS systems like Windows OS and servers centric, Azure makes a reliable, consistent platform between private and public Cloud.clouds. Azure can improve usability and performance via a broader range of hybrid connections, including virtual private networks (VPNs), caches, content delivery networks (CDNs), and ExpressRoute connections.

In 2019, Microsoft-owned LinkedIn moved all its data to the public cloud MS Azure – bringing in the professional information and networks of 645 million members worldwide. LinkedIn is the world's largest professional social media on the Internet that focuses on professional networking and career development.

You can use LinkedIn to find jobs or internships and connect and strengthen professional relationships. LinkedIn also provides the platform for training and learning the skills you need to succeed in your career. As a result, millions of jobs have been fulfilled, and people have been hired by companies or people they connected with on LinkedIn.

Netflix Powered by AWS

A lot of us may subscribe and watch Netflix at home. With 221.64 million subscribers, Netflix has the most extensive subscriber base of all streaming services. Do you know what is the powerhouse behind the Netflix platform to serve such a large customer base? It is Amazon cloud services (AWS). According to Amazon, Netflix sets up and uses more than 100,000 server instances on AWS for nearly all its computing and storage needs, worth about USD$1 billion annually.

This means Netflix uses AWS for almost everything in terms of cloud, including online storage, recommendation engine, video transcoding, databases, and analytics. So, most of the USD$1 billion Netflix plans to spend on cloud services will go to Amazon Cloud Services (AWS).

Netflix's decision to move to the Cloud follows the steps of many large businesses today because of the ever-increasing volumes of data they are handling. With the AWS Cloud, for instance, it is possible for Netflix to quickly and easily scale the company's data warehouse up or down to meet demand.

Moving to the cloud has benefited Netflix in several aspects. It relies on the cloud for its scalable computing and storage needs. This covers hundreds of functions that comprise the Netflix application, such as its business logic, distributed databases, big data processing, analytics, recommendations, and transcoding. AWS also provides Netflix with a secure cloud path for its billing infrastructure and customer and employee data management systems.

In a nutshell, companies like Netflix would be able to rack the servers in their own data centers to handle the ever-growing volumes via its own private cloud. Although such private cloud would let them add virtual servers and large storage quickly as well, the heavy lifting remains on the private cloud management and economy of scale side. Thus migrating to a public cloud like AWS turns out to be a much better choice.

Chapter 10

From Web 1.0 Toward Web 3.0 – The Unified AI Web

World Wide Web is one of the main applications via the Internet. What we routinely use online are mostly from the web.

This book holds the central idea that for successful digital mindset transformation, we must go through some pivotal leaps. This is not just about our knowledge leaps in technologies. This is about the leaps of our cognitive power in assessing and judging technologies as well. As usual, technology has pros and cons. We need to grow the insight and capability to identify technologies with real potential and futures. The following so-called Web 3.0 makes a good example in this regard.

What is the closest daily experience for ordinary readers and users to engage with digitalization? Through three means: PC, smartphone, and the Internet. PC and smartphones are categorized as the end devices in the digital ecosystem. We can see them, hold them, and use them. But the Internet is quite different. Internet means internetworking or a network of networks. It is made of a huge number of sub-networks across the world through a universal protocol called Transport Control Protocol (TCP)/ Internet Protocol (IP) or TCP/IP.

DOI: 10.1201/9781003305187-12

The Internet is the miracle ever happening to digital transformation. Think about it: your smartphone or PC is only an individual or point device, but the Internet can easily link your devices to the rest of the world from anywhere you have a wireline or wireless connection. The Internet is a major enabler of the digital transformation.

The Internet started with the US Department of Defense in the late 1960s when the DoD felt it needed to link individual computers together and allow them to exchange files and communicate. Such a notion soon started to change the world. Universities, businesses, and people at home or even on the road all say they need to be connected too, first via PCs, and now smartphones, tablets, cars, and watches.

Hence no wonder we see more and more sub-networks emerging, which has stimulated the rapid growth of the Internet itself. If you set up a home-based business, buy three PCs, and link them online, you become part of the Internet. You post something like a blog or set up an eCommerce portal online, then people using the Internet across the world can see it. Isn't that amazing? Again, Internet architecture is a meta-network, which refers to thousands of distinct networks interacting with a common protocol, TCP/IP.

TCP/IP is the "world language" for the Internet, so we can keep the Internet open, easy to access, and end-to-end across the globe. Can you imagine the Internet will use a different protocol in each country? Like French in France, English in the UK, and Spanish in Spain? That would be a nightmare for the Internet, right? Do not get it wrong; we can use different languages on the application level or website. But underneath, at the net-working and technical level, it got to be one single protocol shared across the Internet.

Same for Internet routing, which means routing traffic from the United States to China, for instance, a single protocol rule called Border Gateway Protocol (BGP) will be in charge. Only via this approach can the Internet work as the Internet. The network is called LAN or local area network in a local area, like in a business building or campus. The network is called WAN or wide area network across divestment areas, as described in Chapter 8.

The World Wide Web – or WWW, or the Web – is an interconnected system of public web pages accessible via the Internet. Be aware that the Web is not equal to the Internet. Instead, it is part of the Internet and one of the applications built on top of the Internet. Notable examples are wikipedia.org, google.com, and amazon.com. A website can be accessible via a public IP-powered network, such as the Internet.

THE INNOVATIVE LEAP – TOWARD THE UNIFIED WEB

The start of the web functionalizes, such as a basic web page setup, content level, and connectivity, was called Web 1.0. Web 2.0, on the other hand, has brought about the changing trends and is facilitating many activities we do today, such as enhancing creativity, securing information, increasing collaboration, and improving the functionality of Web 1.0.

If you are old enough to recall, Web 1.0 content mainly was "read-only" static Web pages that were not interactive. In other words, you could visit a website to get and read information, but you would not feed it any data back. That is a major difference between Web 1.0 and Web 2.0.

Web 2.0 is about a variety of websites and applications that start to allow people to create and share information or materials online. That created a path toward social media. Web 2.0 makes it possible for web apps, self-publishing platforms like WordPress, and social media sites. We may all have used such Web 2.0 sites as Wikipedia, Facebook, Twitter, and various blogs, and they all have transformed how the same information is shared and delivered.

Now, Web 3.0 is under development to represent the next generation or phase of the evolution of the Web and Internet and could potentially be as disruptive, representing as a big paradigm shift as Web 2.0. But what will Web 3.0 be like remains debatable to date. In 2006, the term Web 3.0 was initially deemed as the third generation of the Internet for websites and applications. The focus would be on using a machine-based understanding of data to provide a data-driven and semantic Web.

Then a new version called Web3 came up in 2014 and has got some attention. Web3 is built upon the core concepts of decentralization, openness, and greater user utility, and it is a democracy in the cyber world on an open-source application. Such a new platform will allow users to control their data and the means entirely to share in the profits generated by their content. Web3 therefore, is a possible future version of the Internet based on blockchains, a record-keeping system best known for facilitating cryptocurrency transactions or distributed finance (DiFi).

There is a thumb-up rule to screen new technologies, i.e., how quickly they can become mainstream, indicating a good adoption rate. Blockchain and cryptocurrency (e.g. bitcoins), however after being touted to the public for years, their adoptions remain limited within some industry circles and there are also growing concerns of fraud and scam from simple decentralization. This means that Web 3.0 remains to be clearly defined. There is another Web 3.0 version that is closer to the notion of Metaverse, and it includes the development of hyper-augmented reality and brain-to-computer interfaces. For instance, implants will be placed over our eyes to access the digital world without needing a display.

The latest trend is, due to the successful global launch of ChatGPT from the end of 2022, generative AI may become a pivotal approach to replace Web 2.0 level technologies represented by Google and other digital giants today. Hence Web 3.0 by the end may very likely be an AI powered and intelligent web.

NEW MINDSET FROM WEB 3.0

Technical Outlook – Universal TCP/IPs

Internet works by using a packet routing network that follows IP and TCP. TCP and IP work together well, ensuring that data transmission across the Internet is consistent and reliable via the device, locations, and access methods.

IP is the rule governing the data format sent via the Internet or local network. An IP address handles the connection between terminals and devices that send and receive information across a network. It uniquely identifies every hardware device on the Internet; without an IP address, there is no way to contact a device.

An IP address functions in two parts: the network ID, which includes the first three numbers in the address, and a host ID which is the fourth number in the address. So, for example, 192.168.1.1, – 192.168.1 makes the network ID, and the final number 1 is the host ID. We typically use four types of IP addresses: a public IP that is limited, used, and exposed in the public network, a private IP that has limitation in capacity and is used in the private domain only, a static IP that is a dedicated and fixed address, and dynamic IP that is flexible and rolling basis with better privacy.

TCP enables the exchange of messages over a network for application programs and computers. TCP is responsible for sending packets across the Internet and ensuring the successful delivery of data and messages over networks. The protocol includes mechanisms to solve many problems arising from packet-based messaging, such as lost packets, out-of-order packets, duplicate packets, and corrupted packets. Hence TCP is used in text communication due to its reliable transmission, error control, and data receiving.

Business Outlook – An AI Website

Artificial Intelligence is changing the world at an unprecedented speed. The development of AI has altered several aspects of our everyday life via the Internet, from knowledge processing to incredibly amazing editing hacks.

Making music or movies or writing an academic thesis used to look like a tiresome, time-consuming endeavor, but not anymore with AI websites. There are numerous AI applications websites that can boost your creativity in a matter of minutes and they have emerged as a result of the debut of numerous AI applications that help you in getting your work done super-fast.

For example, Grammarly, a popular language processing and spelling and grammar checker online tool functions via natural language processing (NLP). The Grammarly team owns deep expertise in NLP, machine learning (ML), and linguistics to create a delightful product for Grammarly's millions of daily active users. The latest is that the firm has integrated GPT-3 into its platform to improve the accuracy and efficiency of its grammar and spelling checks.

Application Outlook – Web Entity and Customer Experiences

As the world moves to digital platforms, customer service can do more and better on the users' experiences. Digital advisors or assistants will replace standalone in-person customer service. These digital reps can now address concerns and queries quickly and 24×7 via web live chat or over the phone. Of course, the service can also be a hybrid of machine and human involvement.

We will begin to see a rise in digital agents. As these agents evolve, they will start to think for us, renew our insurance, and screen advertising messages. The top value for customers who prefer web live chat, for example, is that it lets them get their questions answered immediately. With live chat, you give customers a way to reach you in the exact moment that they have questions or problems they cannot solve. Hence web live chat improves the support experiences of customers.

Users will experience an entirely integrated Internet experience, combining both online and in the real world, with the Internet of Things devices, all communicating and learning via that single digital web identity.

INSIGHTFUL PRACTICE

Nielsen's Law and the Internet

The Internet makes a fantasy world with its own laws. Nielsen's Law of Internet Bandwidth is a crucial one of such laws and states that an average Internet user's connection speed grows by 50% annually. This means the speed doubles roughly every 21 months. Nielsen's Law is important for network design, management, marketing, and services because the highest data speed helps plan and size the network.

The direct reason for such a massive increase in data consumption and demand, as we are all witnessing over the past 60 years, is because of the ever-increasing devices and connections, which are growing faster at 10% CAGR (compound annual growth rate) than both the world population (1% CAGR) and Internet Users (7% CAGR). For example, a connected 4K Ultra High Definition (UHD) TV at your home consumes nearly 15–18 Mbps of data.

Bandwidth is critical for determining how fast a web page loads online. While a fast Internet connection allows you to download web pages and videos flawlessly, higher bandwidth will effectively improve the user experience and let users enjoy the very best from your website. Bandwidth also tells the maximum data transfer rate when you have a network or Internet connection. For example, a gigabit Ethernet via fiber optical connection has a bandwidth of 1,000 Mbps (megabytes per second). An Internet connection via cable modem (copper) may only provide 25 Mbps bandwidth.

Though interrelated, keep in mind that speed and bandwidth are two very different concepts. Network speed is the data transfer rate measurement from a source to a destination system. On the other hand, network bandwidth indicates how much data can be transferred per second or "the size of the pipe."

The telecommunications industry typically uses Nielsen's Law of Internet Bandwidth to represent historical broadband Internet speeds and to forecast future broadband Internet speeds. While Mr. Nielsen predicted many years ago that the high-end user's downstream connection would grow by approximately 50% CAGR, peak service tiers offered by service providers over the years, on the safe side, may follow a servive buffer closer to a 60% CAGR.

Border Gateway Protocol (BGP) – Reigning the Internet Routes

BGP is the routing protocol or traffic driver of the Internet; in other words, it is the protocol enabling the global routing system of the Internet. BGP can connect any internetwork of an autonomous system (AS) using an arbitrary topology and is classified as a path-vector protocol because BGP attributes, including AS numbers, are stored instead of just IP/network hops. BGP is how all Internet routers route our email and web requests across the Internet.

BGP, as a network application, manages how packets get routed from one network to another via the exchange of routing and destination information among edge routers. It makes best-path decisions based on current reachability, hop counts, and other path characteristics. BGP in networking operates on the OSI Transport Layer (Layer 4 – TCP) to control the Network Layer (Layer 3 – IP).

One key feature of BGP is that it offers network stability by guaranteeing routers can quickly adapt and send packets through another reconnection if one Internet path is found going down. The BGP decision-making process analyzes all the data and then decides one of its peers as the next stop before forwarding packets to a certain destination. Each peer hop manages a table with all the routes it knows for each network and propagates that information to its neighboring AS.

BGP is undoubtedly the most complex IP routing protocol currently deployed on the Internet. Its complexity is primarily due to its mission of linking the entire Internet and focuses on security and routing policies.

From a user standpoint, BGP is the magic maker and enables the Internet to route your email and web requests worldwide. Without BGP, no Internet traffic would flow through, and the Internet would be useless.

The challenge with BGP is that the protocol does not directly include security mechanisms and is mainly based on trust between network operators that they will secure their systems correctly and not send incorrect data. But this trust sometimes can get tricky, while BGP was designed only partially to verify the route claims of individual networks. If sometimes certain AS accidentally announces bad routes or is hijacked by hackers to broadcast inaccurate routes, data flows may start to back up or re-route in unexpected ways that can lead to connectivity issues or even outages.

Blockchain May Disrupt Central Banks

Blockchain is the latest buzzword in the marketplace. Although still debatable in its prospect especially from central supervision and fraud prevention angles, a decentralized blockchain network is expected to facilitate Web 3.0, search engines, social media platforms, market-places, etc.

Existing Internet platforms typically host user data and applications on central servers. This allows companies to create detailed profiles of their users. A blockchain-based alternative, on the hand, can change this situation by replacing the centralized topology of the Internet with a decentralized, peer-to-peer network of providers. The blockchain is a completely secure online ledger that records every transaction made at a given place. Blockchain targets to make transactions safer and faster, and the potential impact on eCommerce is tremendous. While the pro is transactin itself may be safer, but the con can be the parties who initailize or handle the transaction may become hard to be authentified, opening doors for some potential bad players and scammers.

Blockchain, therefore, holds a debatable future, or say its advantages are also disadvantages. On one hand it may bring greater efficiency and transparency to the banking industry. For example, it will allow cross-border transactions to be made in real-time and exchange money at Internet speed. On the other hand, the other party overseas may easiliy eacape any safety and secuirty supervision. In a larger sense, blockchain has the opportunity to disrupt the ICT+ banking industry by disintermediating the key services that banks provide, from payments to clearance and settlement systems if meanwhile we find effective ways to mitigate the risks of fraud and scams.

For both brands and buyers in eCommerce, blockchain, when being utilized properly by authentified parties, can also bring about a lot of other benefits. It will help cut costs, improve business processes, make transactions faster, and improve the overall customer experience.

Another big name is Bitcoin, which is the cryptocurrency that led to the blockchain technology. Bitcoin is a virtual currency or a digital currency. It works like an online version of cash. Customers can use BitCoins to make purchases on sites and apps as long as they accept Bitcoin as payment, although not many stores accept Bitcoin yet, and some countries have even banned it altogether. From the turbulent fate of Bitcoins, we can also tell the journey of Web 3.0 remains experimental and challenging.

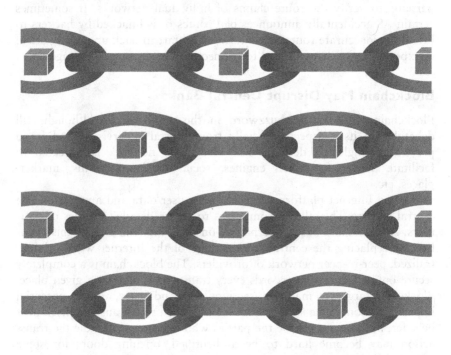

Blockchain is the next-generation network security technology that allows distributed transactions to happen for each individual user with their own privacy.

Chapter 11

From Copper to Fiber Optics – At the Speed of Light

Fiber optics are the glass-made cables that enable high-speed and long-distance data communications.

When we say fiber optics, many people may think it is a technical jargon. But if we say voice, data, and video traversing at the speed of light, nearly

DOI: 10.1201/9781003305187-13

everyone can conceive the meaning. The so-called communication becomes possible only when signals we send go through some medium that carries the signals back and forth. For instance, we can hear each other's voices when we get together and talk. Most people would think this happens naturally and take it for granted.

As a matter of fact, when we talk, the air is the medium because our Earth has air. Sound goes through a medium such as air or water, or specifically through the vibration of atoms and molecules in a medium. In outer space, since there is no air, sound cannot travel, and we would feel all quiet. Hence, we cannot hear any sounds in near-empty regions of outer space.

Today, we use the plural form of medium – media to refer to the press made of TV, social media, online news portal, newspaper, and journals. Why is that? Because they help to transfer information in society. By the same token, we need some medium to carry the traffic through in tele-communications or digital communications. In the beginning, copper was used as phone lines because copper can pass electric signals like an audio voice through and is also not very expensive.

But gradually, copper's constraint started to show up when we needed to pass more traffic, such as data and video, over the circuits and longer distances. Although from the backbone angle, copper cables are technically capable of up to 40 Gbps in speed, from the access point, typically, copper-based Internet speeds reach up only to 300 Mbps. At the same time, fiber optics have been tested to be measurable in hundreds of Terabits per second for backbones and gigabits level for access lines. So compared to fiber, copper becomes minor.

THE INNOVATIVE LEAP – FASTER, LONGER, AND BIGGER BACKBONE

Again, copper-based transmissions max out at 40 Gbps on the network backbone level, which is far from enough for backbone traffic that can quickly go to the terabit level. The solution is fiber optics that can go limitless in bandwidth. An optical fiber is transparent and flexible. It is made by drawing glass (silica) or plastic to a slightly thicker diameter than a human hair.

Light is kept in the core of fiber cable by the principle of total internal reflection, as described in the following details, which causes the fiber to act as a waveguide. While digital data is the new energy source of the information age, optical fiber is the pipeline to carry and distribute this new energy globally.

Nowadays, both wireline and wireless backbones, terrestrial lines, and undersea cables are mostly made of fiber optics. Copper still plays some roles indoors within short distances. Why is that? Although fiber is cheaper than copper, the optical transmission system setup is more expensive than electrical systems. Hence copper can still be more affordable and effective in some indoor applications.

Another factor is latency which is associated with distances and media materials like copper vs. fiber. At longer distances, latency in the fiber-optic system is much lower because of less need for processing and repeating the signals. But if within the size and distance of a house, copper or fiber lines will not make much of a difference in latency.

NEW MINDSET FROM FIBER OPTICS

Technical Outlook – Go Optical

Optical communication follows the principle of total internal reflection. When the injection angle of light meets the conditions required, light can achieve total reflection in the optical fiber, thereby making long-distance transmission possible.

We know copper is metal, while fiber optics is glass. Glass is cheaper also, with much higher transmission speed and bandwidth. That is why fiber is ideal to be used for modern communication media. Optical communications become possible technically when we convert electric signals to light signals.

An optical transmitter is key to take the incoming electrical signal and convert it into an equivalent light signal. The light power is from the laser, which is the device that generates the modulated light signals and transmits them through the optical cable system. The transmission turns each laser device ON to represent a digital 1 and turns it OFF to represent a digital 0.

Recall from Chapter 1, when a signal is generated over a metal like copper, it is an electric and analog signal made of high and low voltages or currents. Fiber-optic cables do not carry any electrical current; they transmit digital binary signals. Hence, we cannot completely replace high-tension copper wires with optical fiber. Electrical and optical signals need to work together: one is good at signal generating, and the other is good at signal transmission.

As for modulated light signals, wavelength-division multiplexing (WDM) is the technique of aggregating and transmitting multiple channels of data through a single optical fiber. The bandwidth of a fiber can be modulated into as many as 160 channels that support a combined bandwidth in the range of 1.6 Tbit/s.

How does it work? The wavelength difference is the key. It sends multiple beams of different wavelengths through the fiber and modulates each beam with a separate information channel. WDM allows the available capacity of optical fibers to be multiplied. The introduction of DWM was the start of optical networking.

Business Outlook – More Fiber, More Data

Optical cables started being used in communication in the 1970s and went booming in the late 1990s with the fast growth of the Internet, generating more and more data communications. By the early 2000s, about 80% of international long-distance data traffic was transmitted over fiber-optic cables. In 1996, the first fully optic fiber cable was deployed underneath the Pacific Ocean, paving the way for faster international data transmission.

Fiber is very future-proof in that it will significantly accommodate the current and future bandwidth increase. It is forecasted that the CAGR - the compound annual growth rate, for the global fiber-optic market will reach 8.5% by 2025 (source: www.statista.com).

Because fiber communication is so reliable and fast, the enterprise could experience speedier access to data and applications that enterprises want to store in the cloud. For example, dark fiber is getting popular for enterprises to use for private networking, such as Internet access or Internet infrastructure networking.

Dark fiber refers to an unused optical fiber system that has been laid but is not currently being used or lighted. Light means a fiber-optic cable transmits data with light pulses; a "dark" cable refers to one through which light pulses are not being transmitted. Dark fiber networks may be point-to-point, or use star, self-healing ring, or mesh topologies.

APPLICATION OUTLOOK – STREAMING, 5G, AND HYPERSCALER

Since fiber-optic cables transmit high capacity of data at very high speeds, they are widely used in Internet cables. Computer networking refers local area network or wide area network in a single building or across campus, or across the country or world and such communications are made easier and faster via fiber-optic cables. Users can see a marked speed-up it takes to transfer files and information across fiber networks.

For 5G buildout, it is essential to note that not only does 5G need fiber as the backbone, but also substantial fiber counts with incredible density and the ability to access that through splice points. 5G requires near enough local breakout to where it is required so the traffic can get picked up.

Fiber communication backbones can go all the way to 800G transmission capabilities to meet the demand from live event streaming, video, gaming, virtual reality, business conferencing, 5G edge applications, cloud computing applications, etc. Copper lines belonged to the age of voice communications, such as making a phone call, while fiber optics take care of hyperscale data communications today.

For instance, while dark fiber can be cost-effective in the long term, its value often lies in hyperscale applications requiring 10 Gbps above throughput, specific routing, and very low latency. Dark fiber bandwidth is virtually infinite and easily adapts to what the business chooses to consume per the applications. Due to the high capital costs in the front with dark fiber, companies should have a full-scale business and workload to justify using and adopting dark fiber.

INSIGHTFUL PRACTICE

Charles Kao Discovered the Purity of Glass

Charles Kuen Kao is well known as the "father of fiber optic communications." He discovered in the 1960s certain physical properties of glass, which laid the groundwork for high-speed data communication in the Information Age.

Back in the mid-1960s, many researchers were focused on using the millimeter waveguide to transmit information and as an alternative to copper wires, and some others were thinking of lasers and fiber optics; Kao combined these ideas, leading tests passing light through different solids to see which may be appropriate candidates for long-distance optical communication.

But before 1966, the dispute was that solids like glass were inappropriate for transmitting signals across great distances. Then Charles K. Kao made a breakthrough in this field. He carefully calculated the model on how to transmit light consistently over long distances via optical glass fibers. Kao's discovery demonstrated that using pure glass could resolve the high loss in fiber-optics communication issue, which proved that optical fibers could be used for high-capacity communications.

Kao pin-pointed the problem that the impurity of the glass materials was the main factor in causing light transmitted to decay. It was not what was believed that there were fundamental problems with fiber optics. He and his colleagues proved that the purest known glass – fused silica – could cut signal loss to the extent of satisfying communications. This paved the basis for fiber communication on a useful scale.

As a result, the first ultrapure fiber was successfully designed and produced just four years later, by 1970. Today optical fibers make up the cabling system that enables our communication society. They are called low-loss glass fibers and facilitate global broadband communication, such as the Internet. As the paise and reward for his discovery of how light can be transmitted through fiber-optic cables, Sir Charles Kao won the Nobel Prize for Physics in 2009.

Fiber Cables Beyond the Sea

A submarine telecommunications cable is laid on the bottom of the sea between landing stations and carries telecommunication signals across the ocean and sea. Undersea cables for transmitting telegraph signals happened even before the telephone invention; the first undersea cable for the telegraph was laid in 1850 between England and France. The Atlantic cable was spanned in 1858 between Ireland and Newfoundland, although its insulation failed and had to be abandoned.

Undersea cables enable instant communications today, transporting about 95% of the data and voice traffic that crosses international boundaries. Undersea cables also form the backbone of the global economy since roughly USD$10 trillion in financial transactions are transmitted via these cables daily.

Submarine cables are laid down by special engineering ships. They carry the submarine cable on board in big reels and slowly lay it out on the seabed based on the plans made by the cable operator. Such vessels can carry with them up to 2,000 km length of cable. Newer ships and plows can now handle about 200 km of cable laying per day.

Today more than 400 subsea cables are in operation around the globe. Some connecting nearby islands can be shorter than 50 miles long. Others traversing the pacific can reach more than 10,000 miles in length. Submarine communications cables transmit 99% of international data. Lines at shallow depths are buried beneath the ocean floor using high-pressure water jets.

Modern submarine cables, of course, use fiber-optic technology. As described in this chapter, lasers on one end of the cable fire at extremely rapid rates along the glass fibers to receptors at the other terminal of the cable. These cables are made of glass fibers wrapped in layers of plastic and, if needed, steel wire for protection. The current world record holder (This may become out of date soon because new cable routes are being laid to the sea every year.) for the longest undersea cable is SEA-ME-WE 3 (SMW3), which stretches 39,000 km (24,233 miles) and connects 33 countries, which would serve an estimated 3 billion people.

Bringing Fiber to the Home

A new study from the Fiber Broadband Association's 2022 Fiber Provider Survey found that there are now a total of 68 million fiber broadband passings in the U.S., up 13% year over year, and up 27% over the past 24 months (source: https://www.cablinginstall.com/cable/article/14288123/fiber-broadband-association-survey-counts-68m-total-ftth-passings-in-us-sharply-trending).

In the past, we used DSL at home for Internet access. DSL is slow because it uses copper phone lines to transmit data. On the other hand, fiber uses ultra-thin glass strands that carry light instead of electricity. We know light can travel very quickly through a fiber-optic cable. Hence fiber connection can see gigabit speeds 100× faster than DSL. Fiber is clearly the way to go and future-proof.

Fiber to the home (FTTH) is much faster and more reliable compared to that ADSL. It delivers a data or voice signal over optical fiber from the operator's switching equipment to a home or business. FTTH is replacing existing copper infrastructure in the homes, such as telephone wires and coaxial cables.

FTTH technology works in two modes – active and passive optical networks, and the latter is also called PONs. The advantage of the PON is that it does not need power and electric switchers. Instead, it can use optical splitters to sort and move data along the network. Therefore, PON can be more cost and operation effective.

There are three basic steps in providing FTTH services: building a backbone network, installing fiber to each block, and connecting homes to the fiber network. The backbone network is typically a fiber ring that feeds the fiber from a distribution node to every block.

With this connection, broadband runs the Internet cable directly to the homes, which helps to offer the fastest Internet speed round the clock with strong connectivity. Advantages of FTTH service providers include very large transmission medium and speed, network immunity against electromagnetic interference, easy and fast installation and low-cost maintenance.

Fiber to the Home (FTTH).

Chapter 12

From Basic Firewall to Cybersecurity – Holistic Cyber Defense

Cybersecurity is the holistic program and architecture to guard and protect end-to-end data use, communications, and storage.

Whenever there is data processing, exchange, or communication happening via whatever media, such as a piece of paper, word of mouth, telephone, telegram, wireless, video, Internet, computer, satellite, cloud, and so on, there is a potential of a security breach of the data and information concerned. No data is secure unless properly secured. In other words, security is the risk and cost of digitalization.

Cyber threats are a big deal. Cyberattacks can cause much damage, such as electrical blackouts, breaches of national security secrets, and failure of military equipment. Security breaches can result in disrupting

DOI: 10.1201/9781003305187-14

phone and computer networks or paralyze systems, making data unavailable, and the theft of valuable, sensitive data like medical records. Top target industries for cyberattacks include healthcare/medical, banking/credit/financial, government/military, education, and energy/utilities.

Conceptually, cybersecurity will remain an ongoing battle. It is a matter of who outsmarts and gets the upper hand between the good guys – regular users and developers and the bad guys, or called the hackers or cyber criminals. What is the prospect of winning for the good guys? We can be conservatively optimistic that good guys will triumph, although the road will be long and hard. We have no choice but to secure cyberspace and our data as best as possible.

Since digitalization is very data centric and enhanced, digital security also becomes a robust industry. What is cyber? It refers to a global domain or the information environment consisting of interdependent information networks, technology infrastructures, and user data, including the Internet, telecommunications networks, computer and cloud systems, embedded processors, and controllers.

As for data security, according to the US Department of Defense, the Five Pillars of Information Assurance model protect confidentiality, integrity, availability, authenticity, and non-repudiation of user data. While the first four pillars speak for themselves, non-repudiation means confirmation and proof that the sender is provided with a proven delivery, and the recipient is provided with the sender's identity. This is to avoid either may later deny having processed the information.

Cybersecurity refers to technologies, processes, and practices designed to protect networks, programs, data, and devices, and prevent attacks, damage, or unauthorized access. Data protection includes sensitive data, personally identifiable information (PII), protected health information (PHI), personal information, intellectual property, data, and governmental and industry information systems.

Cybersecurity is essential because it covers all categories of data, from theft, damage, and other cybercrimes. Here are some major types of cybercrime we frequently face:

- DDoS attacks
- Botnets
- Identity theft
- Cyberstalking
- Social engineering
- Potentially unwanted programs (PUPs)
- Phishing
- Prohibited/illegal content

THE INNOVATIVE LEAP – CYBERSECURITY BEYOND POINT SOLUTIONS

Initially, the IT security solution remained a so-called point solution. Namely, it would focus on and secure a device, a server, a building, or a campus. Traditional and basic IT security practice includes firewall, data encryption, and antivirus software as good examples. The old generation of antivirus software often came in discs for users to install onto their PCs.

A firewall sits between an internal computer network and the Internet to filter out unwanted intrusions. Data encryption is another tool for encoding messages, so it can only be viewed by authorized individuals and is widely used in systems such as e-commerce and Internet banking. Antivirus is a software developed and used to scan, prevent, detect, and delete viruses from a PC or laptop. Once installed, antivirus software runs in the background to provide real-time security and protection against virus attacks.

As digitalization develops, cybersecurity is getting more complex. It has turned into three types: cloud security – cloud-based data storage has become a popular option over the last decade due to its enhanced privacy; network security – guards your internal network against outside threats with increased network security; and application security – like all the malware securities for better application experiences.

There are three major ongoing changes with cybersecurity: first, it becomes from traditional onsite point solutions to cloud-based systematic solutions. Second, it goes beyond traditional physical perimeter-based security to security from anywhere. Third, cybersecurity technologies are becoming more intelligently preventive than mechanically defensive.

Digitalization also brings up a much broader scope and horizon, and the "Big 3" types of cyberattacks cover malware, ransomware, and phishing attacks. Thus the protection scope has become from protecting the location to safeguarding the user, indicating protection nowadays will follow the users wherever they are.

Cybersecurity is hard because it is not just a technical solution. The management of risk is a complex topic that requires significant organizational involvement. This means not only those cybersecurity professionals taking some responsibility for the risk assessment, controls, verification, or recovery but everyone in the organization.

NEW MINDSET FROM CYBERSECURITY

Technical Outlook – Deep Packet Inspection Through All Layers

The three principles of information security, called CIA, refer to confidentiality, integrity, and availability. Every solution of the information security program must implement one or more of these principles.

The good news is that the shield gets more robust as the spear advances. AI and machine learning are major contributors to the advancement of cybersecurity. Machine learning is used to identify hackers' malicious behavior by modeling network behavior and improving overall threat detection.

Network security tools can be either software or hardware based. They help security teams protect their organization's networks, critical infrastructure, and sensitive data from attacks. These include tools such as firewalls, intrusion detection systems, and network-based antivirus programs.

The security approach will also go beyond routine packet checking via a conventional firewall. For instance, deep packet inspection (DPI) makes a sophisticated and advanced method of examining and managing network traffic. It filters packets to locate, identify, classify, and reroute or block packets with specific data or code payloads, while conventional packet filtering examines only packet headers but not other payloads, and cannot detect in depth either.

Business Outlook – Understand Cybersecurity Trends and Principles

As 5G, edge cloud, and other technologies become more common, they will undoubtedly improve our lives but also offer significant security challenges. Three security trends to focus on include 1) the expanding cyberattack surface (remote work, IoT supply chain), 2) Ransomware as a cyber weapon of choice, 3) threats to critical infrastructure via Industrial Control System (ICS), Operation Technology (OT)/Information Technology (IT) cyber-threat convergence.

For an orgnizatin to better protect their systems and data from cyber threats, the cybersecurity principles can provide strategic guidance. We can group these cybersecurity principles into four core pillars, and a well-devised cybersecurity plan should be based on the following four core pillars ensuring data safety:

Pillar I. Identify and manage security risks, and set security policies and planning.

Pillar II. Implement security controls and technologies to reduce security risks.

Pillar III. Understand cybersecurity events, and launch employee education and awareness.

Pillar IV. Respond to, resolve, and recover from cybersecurity incidents via backup and disaster recovery (DR).

By adopting such a security framework planning approach and guidance on solutions, orgnizations can guarantee that any technology they buy will have a unified and shared purpose. The new procurement will be aligned to all other solutions the organization provides and contributes to their ultimate aim of securing their organization's data as effectively and efficiently as possible.

Application Outlook – Cloud-based Security Prevails

Again, cybersecurity entails a series of technologies, processes, policies, and controls that protect systems, networks, devices, data, and programs from cyber risks and attacks. Its goal is to protect against the unauthorized exploitation of systems, networks, and technologies.

Cloud computing is preferred over on-premise servers because of the shared responsibility of security which reduces the workload of organizations. Cloud computing involves only software monitoring, while hardware and software monitoring takes place in the case of on-premise.

While on-premise setups can also keep data secure with high efficiency from the start, as a cloud system learns the enterprise network and grows with it, over time, it can become more secure than on-premise security and become one of the major motivations for organizations to migrate ICT services from on-prem to the cloud.

For example, why is AWS more secure than on-premise? With the cloud on AWS, the security protocols and encryption are built into the entire AWS infrastructure by default. Users pay for computing time and storage space while gaining security and encryption without additional charges.

INSIGHTFUL PRACTICE

The Story of Colonial Pipeline Ransomware Attack

Colonial Pipeline is one of the largest oil pipeline operators in the United States. It was a private business founded in 1962 and HQed in Alpharetta, Georgia. The firm provides roughly 45% of the fuel supply along the US East Coast. Their products include gasoline, diesel, heating oil, jet fuel, home, and military supplies. The company transports over 100 million gallons of fuel each day across an area from Texas to New York.

On May 7, 2021, Colonial Pipeline suffered a ransomware cyberattack that impacted computerized equipment managing the pipeline. The pipeline was the subject of a malware cyberattack. The incident provoked a shutdown of their operations for five days, which resulted in a temporary fuel shortage along the US East Coast.

Ransomware attacks work by gaining access to the victims' computer or device and locking and encrypting the stored data. While a ransom is demanded, there is no guarantee that victims' data will be restored if they pay that ransom. Even if they pay, the attackers may never give the victims the decryption key.

To date, we know few specific details on how the cyberattack took place on Colonial Pipeline's networks maybe until further analysis of the incident. However, we know the ransomware outbreak was linked to a cybercriminal hacking group called the DarkSide, based in Eastern Europe.

The initial attack vector is not publicly known. Still, it may be due to an old, unpatched vulnerability in a system, or it can be a phishing email that successfully fooled an employee. Sometimes hackers can use access credentials purchased or obtained elsewhere that were leaked previously to infiltrate a company's network.

Upon the attack, Colonial Pipeline quickly shutdown the pipeline to prevent the ransomware virus from spreading. A security investigation firm named Mandiant was brought in to investigate the attack. The government agencies, including the FBI, US Department of Homeland Security, Cybersecurity and Infrastructure Security Agency, and US Department of Energy, were also notified of the incident.

Eventually, it was reported that Colonial Pipeline paid the hackers for the decryption key, enabling the company's IT staff to regain control of its systems. Colonial Pipeline restarted pipeline operations on May 12. Then there came a twist on June 7: the US Dept of Justice recovered some USD $2.3 million in cryptocurrency ransom paid by Colonial Pipeline, cracking down on hackers who had launched the most disruptive US cyberattack.

Get to Know DDoS Attacks

DDoS (Distributed denial of service) is a major type of DoS- denial of service attack and can cause major damages. A DDoS attack generates multiple connected online devices, known as a botnet, which can overwhelm a target website with fake traffic and prevent regular traffic from reaching its destination. This can make DDoS extremely destructive to any online organization as it can make an online service unavailable and take the network down.

On November 15, 2021, cloud infrastructure company Cloudflare said it had blocked a DDoS attack that peaked at just under 2 Tbps, one of the largest DDoS ever recorded. In February 2020, AWS reported preventing and mitigating a severe DDoS attack targeted at an undisclosed AWS key customer. The attack lasted for three days at the massive level of 2.3 Tbps. Although the AWS DDoS attack was quickly mitigated and didn't cause much damage, still, that magnitude of DDoS targeting a key customer of a giant cloud computing provider like AWS made cybersecurity experts very nervous.

The first documented DoS-style attack occurred during the week of February 7, 2000, when "mafia boy," a 15-year-old Canadian hacker, orchestrated a series of DoS attacks against several e-commerce sites, including Amazon and eBay. In recent years, we see an exponential increase in DDoS attacks that have crippled businesses for significant amounts of

DDoS attack is one of the major hacking methods to breach network security and obstruct regular functions of the network and servers, by overflowing traffic to the destination targets.

time. *InfoSecurity Magazine* (source: https://www.infosecurity-magazine.com/) reported 2.9 million DDoS attacks in Q1 of 2021, an increase of 31% over the same period in 2020.

Usually, the goal of a DDoS attack is to overwhelm the website resources and cripple the use of the website. Nevertheless, DDoS attacks can also be for extortion and blackmailing. In such cases, website owners can be extorted for a ransom for attackers to stop or avoid a DDoS attack. To protect yourself from DDoS attacks, equip your network, applications, and infrastructure with multi-level protection strategies. This may include pre-vention management systems that combine firewalls, VPNs, anti-spam, content filtering, and other security layers to monitor activities and identify traffic inconsistencies that may be symptoms of DDoS attacks.

Evolution of Antivirus Software for PC and Web

As described above, an antivirus product is a software program to detect, mitigate, and remove viruses and other malicious software from PCs or laptops. The users must always install antivirus software and keep the version up to date to protect their data and devices.

Many PC users have recently found that once they installed Windows 10, they seem not in need of antivirus anymore. To be clear, antivirus is always necessary, but most people do not need an extra layer of protection. The built-in antivirus in Windows 10 can do a good job in your PC or Mac. The new version of Microsoft's Defender, for instance, is very effective at detecting malware files, blocking exploits and network-based attacks, and flagging phishing sites. It includes simple PC performance, health reports, and parental controls with content filtering, usage limitations, and location tracking.

Microsoft Defender was first introduced with Windows XP as a free tool to protect MS Windows users from viruses, malware, and other spyware. Now it helps detect and remove viruses like the Trojan from your Windows 10 system. As mentioned above, Windows 10 has built-in antivirus pro-tection, which is Windows Defender, but sometimes it may need additional software, either Defender for Endpoint or a third-party antivirus.

Hence you may use Windows Defender Anti-Malware, Windows Firewall, or McAfee Anti-Malware and McAfee Firewall. But no need to install both; if you use Windows Defender, then you can completely remove McAfee. If you are using Windows 11 in S-mode, your PC and data will be secure in most parts, with no need to buy an antivirus any longer.

From the web browser perspective, Microsoft Edge is more secure than Google Chrome for your business on Windows 10. Once you have MS Edge in use, no additional software is required to achieve the secure baseline. The browser has robust, built-in defenses against phishing and malware and natively supports hardware isolation on Windows 10.

Chapter 13

From Onsite Database to Cloud Storage – Data Sustainability

Cloud storage is to migrate the traditional hard disk storage to the cloud with more capacity and agility, and cost-effectiveness.

Common sense is that when we collect and process digital data, we need a reliable, accessible, and secure repository to keep the data or call it data storage. Most of us know about data storage approaches and means such as discs, tapes, database servers, and data centers.

Digital information has two types: input and output data. Users provide the input data, and computers provide the output data. Input can also be called data entry: users can input the data directly into a computer. However, we know continual data entry manually can be time and energy prohibitive.

One short-term solution is using computer memory, also known as random access memory (RAM). But RAM's capacity and memory retention are limited. Another means is read-only memory (ROM); as the name says, the data can only be read but not edited. Advances in computer memory, such as dynamic RAM (DRAM) and synchronous DRAM (SDRAM), are still limited by cost, space, and memory retention. When a computer powers

DOI: 10.1201/9781003305187-15

down, RAM is down and gone too. The solution to this? secondary data storage.

Most of us are familiar with two types of computer storage devices: RAM is an example of a primary device, and the hard drive is an example of a secondary device which is more robust. Secondary storage, as we often use, can be removable, internal, or external. The internal is called direct area storage, and the external is called network-based storage, i.e., cloud-based storage.

Data storage is rarely a problem at the individual level. Most of us use our PC, a USB disk, or external data disk drive to store our data, which is mostly sufficient. When smartphones came up, however, the storage of data became cloud based, like iPhone using iCloud; otherwise, apparently, there would not be enough space for a small device like iPhone to handle much data.

On the other hand, business data storage is more complicated in terms of process, technology, and architecture. Before cloud computing came along, data storage had challenged many big organizations. Many companies had to resort to magnetic, optical, or mechanical media that record and preserve digital information for ongoing or future operations.

THE INNOVATIVE LEAP – CLOUD STORAGE AS A SERVICE

The most important part of data storage is apparently about its capacity, namely, how much disk space one or more storage devices can provide, which measures how much data a computer system may contain. For example, a computer with a 500 GB hard drive means it can store up to 500 GB. But from an individual device angle, such storage capacity is always limited.

Cloud storage as a cloud computing model emerges to make a big difference. It stores data via the cloud through a cloud computing provider (CSP) that manages, operates, and secures data storage as a service. It is delivered on-demand with scalable capacity and prices, which saves the cost and resources of purchasing and opreating your own data storage infrastructure. Compared to an external hard drive, cloud storage and backup technology also win easily in terms of accessibility; that is to say to access cloud storage, all you need is a working device with a reliable broadband Internet connection.

Here is a quick summary of cloud-based storage features:

- Storage capacity and scalability on demand
- Usability and accessibility
- Better security
- Cost-efficient
- Convenient sharing of files with multiple users
- Automation and synchronization

Some companies may use a hybrid storage approach using both hard drives and the cloud: in general, external hard drives provide a good and convenient medium for your backup data. Then in case of physical threats such as fire, flooding, or other dangers, the secure cloud storage services provide your data an extra layer of protection by storing your important files offsite.

NEW MINDSET FROM CLOUD DATA STORAGE

Technical Outlook – Data Storage Formats

By instructing computers to access and store data from the storage devices, we use computers, PCs, or terminals to connect to storage devices directly or through a network. Generally speaking, data storage has two key parts: the format and structure on which data takes on and the devices where data is recorded and stored.

In digital time, data can be recorded and stored in three primary formats: file storage, block storage, and object storage. File storage is a hierarchical methodology we frequently use to organize and store data. We also call them file-level or file-based storage to store data in files organized in folders. The folders are organized under a framework of directories and subdirectories.

Block storage, also called block-level storage, is a technology that stores data into blocks, and each storage unit acts as an individual hard drive (e.g., in our PC, we can have hard drives C, E, F, etc.) that the storage administrator configures. Since the blocks are stored as separate units, each with a unique identifier, IT Developers typically favor block storage for computing projects that require fast, efficient, and reliable data transfer.

Object storage, or object-based storage usually handles large amounts of unstructured data that does not conform to, or is hard to be organized into, traditional files or relational database format. Examples include email, videos, web pages, audio files, photos, sensor data, and other media and web content (textual or non-textual).

As for the devices to store data, direct area storage, also called as direct-attached storage (DAS), is what we mostly use daily. This storage is often directly connected to the computing machine locally. DAS can provide decent local backup services, but data sharing is limited. We have seen a lot of DAS devices, such as floppy disks and optical discs, including CDs (compact discs), DVDs (digital video discs), HDDs (hard disk drives), flash drives, and SSD (solid-state drives).

On the other hand, network-based storage allows multi- computers to access it via a network, making it easier for data sharing and collaboration. Its remote and off-site storage capability is ideal for backups and data protection. There are two common network-based storage setups: network-attached storage (NAS) and storage area network (SAN).

Business Outlook – Hyperscale Capacity OnDemand

Now data storage is in high demand for high-level computational needs such as big data analytics, AI (artificial intelligence), ML (machine learning), and IoT (the Internet of things). And the other demand for huge data storage amounts is protection against data loss in disaster, failure, or fraud. More organizations also employ data storage as backup solutions to avoid data loss.

As storage demand soars, tech advances and businesses shift toward cloud storage, ICT vendors can expect double-digit adoption growth for technologies such as high-capacity hard disk drives, persistent memory, all-flash storage arrays, cloud storage infrastructure, and file-sharing services.

Referring to a technology roadmap put forward by the Advanced Storage Technology Consortium (https://astc-inc.com/), the capacity of HDDs (Hard Disc Drives) will rise to 100 TB by 2025. New writing technologies will be enabled, such as Shingled Magnetic Recording, Perpendicular Magnetic Recording, Enhanced Caching, and even helium inside the casing.

What is Helium? Helium is a chemical element used for storage to prevent overheated casing drives due to huge amounts of data. Helium has its unique value in its cooling properties because it has a boiling point close to absolute zero.

Application Outlook – BaaS and Hybrid Cloud Storage

Backup storage and appliances target to protect data loss from disaster, failure, or fraud. The way it works is to periodically copy data and application to a separate, secondary device. Backup appliances range from HDDs and solid-state disks (SSDs) to tape drives to servers.

Now cloud can also offer backup storage service, namely, backup-as-a-service (BaaS). As most as-a-service solutions offer, BaaS provides a low-cost option for data protection from a remote location with scalability. The cloud service provider manages and maintains the servers and associated infrastructure. The users have access to the data whenever they need it.

A popular alternative is hybrid cloud storage that combines private and public cloud elements. Hybrid cloud storage allows organizations to choose which cloud to store data. For instance, highly regulated data with strict archiving and replication requirements are better suited to a private cloud environment. On the other hand, less sensitive and risky data can be stored in the public cloud. Some organizations also use hybrid clouds to supplement their in-house storage networks with public cloud storage.

INSIGHTFUL PRACTICE

iCloud Makes a Difference

We know every user with an Apple ID gets 5 GB of storage for free, and the storage is used to backup apps and settings from their iPhone, iPad, or Apple watch. With more benefits, the service also syncs your photos, documents, and emails with your Mac and mobile devices. Again, the idea behind iCloud is to be able to access all your stuff no matter what Apple device you are using.

iCloud therefore as a sweet offer from Apple, is designed to securely store users' photos, notes, passwords, files, and other data in the cloud. It can automatically keep the data up to date across all your devices. iCloud also makes it easy to share files, notes, photos, and more with friends and family. Can you imagine your iPhone would work well without iCloud?

The purpose of iCloud is to securely store data and important information on Apple's remote servers in the cloud since your iPhone or iPad is too small to handle such storage locally. This way, all your data is backed up to a secure remote location and synchronized between your devices, and these are the two major benefits of having your information backed up to the cloud.

What about the content you have bought? Your iCloud Backup can include information about the content you buy rather than the content itself. When you restore purchased content from an iCloud backup, it is automatically redownloaded from the iTunes Store, App Store, or Books Store, although some types of content are not downloaded automatically in all countries or regions.

When the content is in transit, iCloud secures your information by encrypting it and storing it in an encrypted format. Many Apple services use end-to-end encryption, meaning only you can access your information and only on trusted devices where you are signed in with your Apple ID. Encryption keys are stored on Apple's servers.

Disaster Recovery (DR) via Cloud

DR is an organization's means of regaining access and functionality to its IT infrastructure, usually after a natural disaster, cyber-attack, or even business disruptions related to technical glitches. A DR plan is made of a variety of DR methods.

The DR architecture is pivotal, and resources are usually duplicated and replicated to a different data center and geographical region to ensure that they are far away from the disaster's source. Today cloud DR is the central function of any business continuity (BC) strategy and is virtually identical to traditional DR in terms of objective: protecting valuable business resources. It also ensures that protected resources can be accessed, managed, and recovered to continue normal business operations.

Data center storage can supplement onsite and cloud storage as a disaster recovering means.

Plus, cloud-based DR is much faster than on-premises DR and does not require as much complexity. This simplicity also allows for easy testing of the DR services, so your company can ensure your DR plans are fully functional.

Businesses use cloud DR to minimize the overall impact of a disaster on business performance, which is vital for a DR site to resume business operations. Cloud backup has the data and applications on a business's servers backed up and stored on a remote server. Customers typically manage the backup and restore their data and apps using a web dashboard or a service provider's control panel.

There is another related concept called high availability (HA), and it is worth to note DR and HA focus on different problems. DR is a storage technology, while HA is technology-based, mostly on networking topology with redundancy. HA systems follow the same "hot and active" backups as DR. In essence, DR picks up when HA fails.

Is Your Firm Using AWS S3?

As businesses use more cloud-based storage nowadays, the most popular cloud-based storage service out there is Amazon Simple Storage Service (or AWS S3), which offers object storage with good infrastructure and speeds for businesses. Storing data as objects in buckets, AWS S3 is accessible for storing and retrieving large capacities of data at any time from anywhere at affordable cost.

AWS S3 can have many use cases, including:

- Storage for the Internet
- Backup and disaster recovery
- Analytics
- Data archiving
- Static website hosting
- Security and compliance

An AWS S3 bucket name is globally unique and secure, and the namespace does not allow duplications. After a bucket is created, the bucket's name will stay unique and cannot be used by another AWS account in any AWS Region until the bucket is deleted.

From a proof of concept angle, AWS S3 provides a free tier to get you started, with a limited capacity for 12 months. S3 is popular, especially for small and mid-size businesses, and the service is well known for its easy accessibility, security, and scalability, where multiple users can read or write data simultaneously with valid authentication.

Chapter 14

From Server Boxes to Virtualization – Catalyst of Cloud Computing

Virtualization is to use software to simulate hardware or OS via hypervisors and make an effective way to run ICT operations.

Maybe we've heard a lot about cloud computing nowadays. What really makes cloud computing possible? Cloud is hosted from data centers, but how is the cloud different from a regular data center service? Here is the magic buzzword that makes the difference: virtualization.

How does cloud computing get the cloud name? Because it works like a cloud floating big and high out there, but with no hard substance within. It's all virtual. It is like if you fly into a cloud, you hit nothing hard there, although the cloud does have its climate functions. By the same token, cloud computing is virtual but functions as the next-gen infrastructure for all the ICT services we need today.

Virtualization, by nature, is a software simulation. In the past, people got used to an IT function that got procured and provisioned as a server box.

DOI: 10.1201/9781003305187-16

One box for the email server, one box for the firewall, one box for the web server, one box for the router, and so on. By the end, our IT operation room was full of a myriad of boxes for point solutions.

Such an IT server box is essentially made of hardware and software, with some accessories called middleware too. Hence, the in-box thinking is that each major software application comes with a hardware frame. However, the out-of-box thinking is, why not we put multiple software over one hardware frame? This way we can save on hardware and make service functions more integrated, scalable, and easier to manage.

THE INNOVATIVE LEAP – GOING VIRTUAL ICT

A brilliant idea it is, but virtualization is not really new. For instance, on our PC that we have used for many years at home, we load various types of software into it already, right? But remember that these home PC applications are typically pretty small in scale and, therefore, much easier to manage and operate.

On the other hand, to put multiple used-to-be business class stand-alone software functions now over a single hardware box, we need to resort to virtualization, which includes two key technologies: hypervisor and virtual machine (VM) to make it work.

A hypervisor, or VMM (VM monitor), is a software program that functions like a "bridge" – it runs on an actual host hardware platform under and governs the guest operating systems (OSs) on the VMs above. A hypervisor separates an OS and applications from the underlying physical hardware and allows the hardware host machine to operate multiple VMs as guests. Such separations or decoupling help "maximize" (This is the key word here) the effective use of computing resources such as memory, CPU cycles, and network bandwidth.

Virtualization is a set of software technologies that enable applications to run on virtual hardware via VMs and hypervisors or virtual OSs via containers (also refer to Chapter 18). A VM, also known as a guest machine, therefore is a software simulation of a hardware platform. It provides a virtual operating environment for guest OSs. When they are maximized, that means we can create a lot of VMs virtually.

With virtualization, we still have a firewall, an email server, a web server, etc., with no changes in their functionalities. The only change is that now they are in the form of software from the cloud, no longer a physical box. When virtualization duplicates dozens of or hundreds of VMs over single hardware, it can make a huge impact on IT operation in terms of effectiveness, efficiency, cost saving, control and management, upgrading, etc.; thus, it becomes a "cloud" that implies scalability, agility, availability, easy access, and rapid provisioning in ICT operations.

NEW MINDSET FROM VIRTUALIZATION

Technical Outlook – Software Simulation

Virtualization uses software to simulate hardware functionality, and this is revolutionary. This creates a virtual computer system and enables ICT organizations to run multiple virtual systems, including OSs and applications, over a single server.

A VM is a simulated computer system that runs on a physical computer. In other words, a VM is a computer inside a computer. VMs allocate memory, a virtual CPU, disk storage space, and a network interface. When selecting virtualization hardware, you have three elements: CPU, memory, and network I/O capacity. They are all critical for workload consolidation.

VMs share memory with the Hyper-V host, so you must have enough memory capacity to handle the expected virtual workload. This means the relationship between the VMs and underlying hardware is that, for instance, if you are running a 16 GB RAM computer with three VMs, then the 16 GB of RAM is divided evenly among all four systems (including the computer itself as hardware). You will not have performance issues as long as none of the systems use more than 4 GB RAM.

Business Outlook – One to Many Efficiencies

Virtualization reduces hardware, helps lower energy costs, and lessens a company's carbon dioxide (CO_2) emissions. Meanwhile, it increases efficiency and business continuity in addition to cost savings. Virtualization has other benefits, including improving staff productivity, business continuity, and disaster recovery.

There are many reasons for companies to start to consider using VMs. If for cost reduction purposes, VMs allow for reduced overhead, with multiple systems simultaneously operating from the same console. If for better management and security purposes, VMs, as another value added, also provide a safety buffer for your data, as they can enable rapid disaster recovery and automatic backups.

One may ask, how many VMs can run over a single hardware host like x86–64? Well, if we use all the processors, we can run up to 64 VMs with stable performance. Isn't that amazing? Think that there used to be 64 individual box devices in the room. Now we can run them just over one single box.

Application Outlook – Virtualization of Everything

Four major virtualization areas can cover every ITC functionality: network virtualization, storage virtualization; desktop virtualization; and application virtualization.

Network virtualization takes the available hardware resources on a network and boosts bandwidth management via virtual software instances. Network function virtualization, or NFV, refers to reducing cost and accelerating device service deployment. Network operators can decouple network functions such as a firewall or data encryption from dedicated hardware and move them to virtual servers.

The goal of NFV is to transform the way that network operators architect networks. It uses virtualization technology to consolidate many network equipment types onto industry-standard high-volume servers, switches, and storage. These devices otherwise may become too many and overwhelm the data center and the network.

In storage virtualization (refer to Chapter 13), multiple physical disks are combined into a group. From that group of physical disks, virtual storage or logical storage blocks are assigned to a server for use or a pool of available storage capacity managed from a central console. There are three advantages to implementing storage virtualization: improved storage management in a heterogeneous IT environment, better availability, estimation of downtime with automated management, and better storage utilization.

Application virtualization means running an application on a machine that does not have the application installed. Instead, the application resides on a VM on a server in a different location and uses the OS of that remote server. An example is application and desktop virtualization, which enables centralized management of the complete desktop environment ecosystem.

Such applications and virtualized desktops can deploy updates consistently, completely, and rapidly. People now are running their desktop only via a local hardware, while all the software and apps are virtually and remotely from the cloud.

INSIGHTFUL PRACTICE

The Magic of Virtual Machines (VMs)

A VM is a computer file, typically called an image, that behaves like an actual computer, or can say it is a compute resource in software form instead of physical hardware to run programs and deploy apps.

The physical machine hosts the VMs running on the top as the guests. Again, this process is managed by a hypervisor. Each VM runs its OS and functions independently from the other VMs, even though they share the same hardware host. VM separates OS on a single computer, and each OS has a share of the computer's system resources. Other benefits of VMs include easy provisioning and maintainability, and high availability.

From a user angle, VMs allow users to run an OS in a separate app window on their desktop that behaves like a full, separate computer. Users can play around VMs with different OSs, run software their main OS cannot support, and try out apps in a safe, sandboxed environment.

Cloud computing becomes possible because it runs multiple virtual machines over a single bare metal hardware.

A system VM is fully virtualized and can substitute for a physical machine. We can take an example of a process VM like the Java VM. It enables any OS to run Java applications as if they were native to that system. VMs also fit well in testing different configurations and setups. IT developers can use VM snapshots to try various scenarios without modifying the hardware architecture.

In the VM environment, developers and software testers identify configuration issues and deficiencies in a very cost and time-effective way before end-users run into them. The testers can turn on as many VMs as they need for the project without any hardware cost. Once the test is done, the VMs can be easily removed or shut down.

Virtual Desktop Deepening Your PC

A virtual desktop offers users to access their desktop and applications anywhere on any endpoint device via a network. IT organizations usually deploy and manage these desktops from a data center or cloud.

To play virtual desktop, you will still need a PC with a minimum CPU, like an Intel Core i5-2500K. The critical part is the minimum memory requirement for a virtual desktop, around 4 GB of RAM installed on your computer. Virtual desktop will run on PC systems with Windows 7 SP1, Windows 8.1, Windows 10, or Windows 11 and upwards.

Now MS Windows 10 has added virtual desktops as a built-in feature. Virtual desktops offer a convenient way to stay organized and optimized if you have a lot of apps open at once or use your PC for different tasks. To quickly launch a new virtual desktop, press Windows+Ctrl+D on the keyboard at any time, and you will immediately be taken to the new desktop. Alternatively, you can click "New Desktop" in Task View. Windows 11 can do even better and let you create custom arrangements of apps on virtual desktops. You can switch the apps quickly using the Task View button.

From a security perspective, the virtual desktop is designed to stream virtual computing desktops to nearly any PC or smart device. One advantage of VDI (virtual desktop infrastructure) is that, the user's hardware compartmentalizes everything performed within this virtual computing environment. This makes virtual desktops much more secure from the perspective of data loss prevention.

Virtual CPE – Consolidating the Boxes

Customer premises equipment (CPE) refers to the equipment employed on a client's (other than a network carrier's) premises to originate, route, or terminate telecommunications, including switches, routers, PBX, email servers, firewalls, etc. As previously mentioned, a regular CPE mostly came as a box or, say, a piece of equipment or device.

Virtual CPE (vCPE) is a new way to deliver network services such as routing, firewall security, and virtual private network connectivity to enterprises. It uses software rather than dedicated hardware devices. The software runs on top of simple, inexpensive, and on-site hardware, which typically uses industry-standard x86 devices rather than function-specific appliances. The cost-effective standard x86 servers can operate such enterprise network edge functions as WAN edge routers, WAN optimization controllers (WOCs), and security functions like firewalls.

Universal CPE (uCPE) is an enhanced version of vCPEs, solving some limitations. uCPE provides a general-purpose platform or a commodity, off-the-shelf server that integrates computing, storage, and networking, allowing it to provide multiple VNFs at customer locations. As VNFs run on uCPE, the configuration and running of network services can be centralized.

In a nutshell, the benefits of using vCPE or uCPE include the following:

- Reduced Capex when buying less equipment
- Reduced OPEX with a simpler network setup
- The ability to run more applications on fewer appliances
- Improved adaptability by easily changing the network
- Centralized appliance management

Layer 3 – Smart Digital Behaviors and Capabilities – Sense, Think, Connect, Communicate, Visualize, and Act

Layer 3 of the Cognitive Model of Digital Transformation describes in detail how digitalization can do things differently than traditional ways. If we say Layers 1 and 2 prepare the digital mindset leaps more from the conceptual and core solution angles, then Layer 3 is starting the immersing stage of the digital journey. Namely, we gain more functional insights into what digitalization can accomplish in all major fields and how it extends and innovates our human capabilities to the next level and new horizons. Together with intelligent machines and apps, we can make great miracles happen in the digital era.

> If you really think about it, we've gone from a fairly episodic approach every once in a while – kind of coming up with the next big idea – to really persistent modernization. That's where we've always wanted to be.
>
> – Lt. Gen. Thomas Todd, the Chief Innovation Officer,
> US Army Futures Command

> Every industry and every organization will have to transform itself in the next few years. What is coming at us is bigger than the original Internet, and you need to understand it, get on board with it, and figure out how to transform your business.
>
> – Tim O'Reilly, Founder & CEO of O'Reilly Media

DOI: 10.1201/9781003305187-17

Layer 3 – Smart Digital Behaviors and Capabilities Sense, Think, Connect, Communicate, Visualize, and Act

Chapter 15

From Online People to Things – IoT Leading Digital Ecosystem

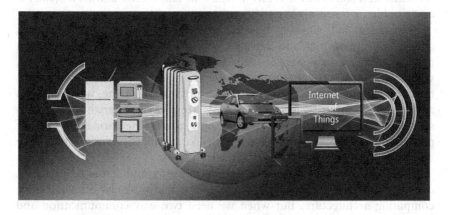

IoT refers to the Internet of things that uses digital technologies like the Internet to connect devices and sensors and makes them online managed services.

When humans get digital in doing things, we often take it as we are using digital tools. But when things and devices around us, such as bank ATMs, cars, home door locks, and industrial environmental monitoring systems also get digital, we are immersing into a new digital world called IoT (Internet of Things) that is generating a digital ecosystem.

The IoT devices worldwide are forecast to almost triple the volume, from 2020 about 8.74 billion to more than 25.4 billion IoT devices in 2030 (source: www.statista.com).

Some people may argue, is IoT really a new thing? Starting from day 1 of the Internet, we are connected by things, i.e., devices such as PCs, laptops, smartphones, and tablets. Well, the difference is that human is using these devices, while for IoT, these devices would fare by themselves or are called self-service.

Like an ATM from a bank, it will sit in some public places by itself without having an administrator standing by. In the past, there was no real-time monitoring of how these ATMs worked, their inventory level, and any

DOI: 10.1201/9781003305187-18

potential troubles that might pop up. They could only be checked on periodically by personnel dispatched by the bank to the site. So, the whole ATM operation was in a semi-dumb mode.

THE INNOVATIVE LEAP – DIGITAL DEVICE SELF SERVICE

Now IoT is changing all of this. IoT means all devices can now be connected via a network, controlled, and managed remotely from the cloud. By embedding monitors, sensors, chips, and communication modules into the ATM, no matter where they are located, banks can now track the performance and status of the ATM in real-time and receive alarms instantly if any trouble or fraud is coming up.

The general IoT working flow has four parts. The front end includes devices with monitors and sensors, embedded chips, and robotics; the network part, mostly wireless like LTE or 5G, sometimes via satellite, which will handle communications between the devices and control center; the control center or user interface, either physical or online, where all data and signals received from the front ends are processed, analyzed, and responded; the cloud storage and big database, this is where some data will be processed in depth and come back with like intelligence and performance reports and proper responses.

If the IoT is mostly about one-way monitoring and status reporting, it can be handled via the so-called low power, low bandwidth, and edge computing architecture. But when we need two-way communication and work happening between the front end and back end, for example, remote surgery, driverless cars, agriculture irrigation system management, etc., that level of IoT would require high energy, bandwidth, and hybrid computing to function right.

This would mean, technically, we need highly efficient front devices, private 5G level and above bandwidth and throughput, and a hybrid of edge and central cloud computing and data processing power to make the high-end IoT work seamlessly.

NEW MINDSET FROM IOT

Technical Outlook – Connecting to the Edge

IoT connects all potential objects to interact with each other on the Internet to provide humans with a secure and comfortable life. As IoT makes our world as possible as connected, embedded computing devices would be exposed to Internet influence.

The general operation of an IoT application includes four stages; data acquisition, data processing, data storage, and data transmission. The first and last stages go on every application, but the processing and storage may or may not run in some applications.

The potential of IoT will only materialize with the ability to use AI alone or supported by human intelligence collection. This will aggregate and analyze the enormous amount of data created by IoT devices. The future and potential of IoT are limitless. We can expect the advances in the industrial Internet will be accelerated. These key factors include integrated artificial intelligence (AI), increased network agility, and the capacity to deploy, automate, orchestrate, and secure diverse use cases in hyperscale.

Business Outlook – Building the Digital Ecosystem

The IoT, together with cloud technology, AI, and machine learning, has been one of the major drivers in digital high-tech over the past couple of years. It has been developing at fast speeds since its inception, often popping up in new and quite unexpected apps.

The benefits of the IoT to citizens, businesses, and governments will be significant. It ranges from helping reduce healthcare costs, improving quality of life, reducing carbon footprints, increasing access to education for remote, underserved communities, and improving transportation safety.

Concerns about IoT include security and privacy. Once a device gets connected or online, hackers can manage to breach the connection and steal the data, or devices at home can connect what you deem as private data that may get exposed to 3rd parties. These are typical pros and cons of the new digital world that we will enjoy and cope with.

Application Outlook – All Verticals Use IoT

Typical IoT benefits include intelligence, automation, and remote and centralized management. Typical IoT applications include smart home, smart city, rural or marine remote operations, and fleet management. Generally, IoT is mostly in use in manufacturing, transportation, and utility organizations via sensors and other IoT devices. IoT has also found use cases for organizations within the agriculture, infrastructure, and home automation industries, leading some organizations toward digital transformation.

Applications run on IoT devices and can be explicitly applied to almost every industry and vertical, including healthcare, smart homes and buildings, automotive, industrial automation, and wearable technology. IoT applications increasingly use AI and machine learning to add intelligence to devices.

For instance, thanks to IoT, doctors can now easily monitor patients at home and better manage their prescriptions by getting vital information like blood sugar and blood pressure. This means they can provide better comfort and safety to their patients, especially when time is of the essence.

INSIGHTFUL PRACTICE

The Smart City Era

A smart city means to make the urban places where we live and work smart. A smart city can improve operational efficiency and share information with the public via information and communication technology (ICT), providing a better quality of government service and citizen welfare.

The infrastructure core elements in a smart city would include modern capabilities such as adequate water supply, solid waste management, assured electricity supply, sanitation, public transport, efficient urban mobility, affordable housing, robust IT connectivity and digitalization, and good governance.

IoT devices come from connected sensors, lights, and meters and collect and analyze data. The cities then process the data to improve infrastructure, public utilities and services, and more. Smart cities are powered by the development of technologies such as IoT, AI, Blockchain, or Geospatial Technology. All this propels the growth of smart cities around the world.

Sensing is at the heart of a smart city's infrastructures, which can monitor things and act on their own intelligently. Sensors provide awareness that enables more efficient use of resources based on the data collected by these sensors, measuring the physical properties of any object or as a bridge, road, building, or situation such as traffic control, water quality, or disaster alarms. The main sensors include biosensors and electronic, chemical, and smart grid sensors.

Smart city is to apply modern ITC a digital technology to better manage and administrate the complex operations of modern cities.

What is the difference this makes from traditional ways of city management? Smart cities bring about better use of space, less traffic, cleaner air, and more efficient civic services. All of this increases the quality of life. Further, smart cities provide more career and economic opportunities and stronger links with the community. A smart city will also help its citizens make decisions on investment, food and restaurant, and hospitals. Measuring visitor and citizen traffic across the city will help entrepreneurs to obtain data that point to where to open new businesses such as restaurants, retail stores, and dry cleaning.

Smart Agriculture and Farming

A digital farm is not an imagination anymore. Smart farming is a new management concept providing the agricultural vertical with the new infrastructure to leverage advanced technology – including the cloud, IoT apps, and big data – for tracking, monitoring, automating, and analyzing operations.

IoT smart farming solutions forge a system that can monitor the crop field with the help of sensors, including light, humidity, soil moisture, crop health, chemical application, temperature, dam levels, livestock health, etc. The system monitors vehicles and weather and automates the irrigation system. Thus, IoT enables devices across farmland to monitor and measure all kinds of data remotely, which can be provided to the farmer anywhere in real-time.

Smart agriculture aims to achieve three main objectives: increasing agricultural productivity, incomes, and sustaining; reducing and removing greenhouse gas emissions; and adapting and building resilience to climate change and natural events, where possible and as much as possible. It will help the decision-making process for better farm management and the preservation of resources. This intelligent system optimizes and examines how small- to large-scale farming can aid the production output.

Smart agriculture can also replace scarce labor in rural areas and enable risk management in the production and distribution process. For example, farmers can now use drones, geolocators, and sensors to improve their farming practices. In this way, it supports sustainable and cost-effective agriculture via navigation satellites and earth observation input, making it easy for farmers to make informed decisions when farming.

Smart Grid for Smart Power

Smart grid is one of the most important applications of IoT, making a two-way flow of electricity and data possible with digital communications technology enabling detection and reaction. Integrated with the power grid to collect and analyze data, the smart grids have sources of data from transmission lines, distribution substations, and consumers.

Comprising computers, automation, controls, and new technologies and equipment, a smart grid represents an unprecedented opportunity to lift the energy industry onto a new level of reliability, efficiency, and availability. For instances, smart grids can be proactive to changes in usage and multiple issues and have self-healing capabilities, allowing electricity customers to become active participants.

By bringing computer technology to a standard electricity grid, now samrt grids allow easier communication between energy retailers, distributors, and customers. Smart grids are considered "smart" because it entails self-healing: fault protection, outage management, dynamic control of voltage, weather data integration, centralized capacitor bank control, distribution and substation automation, advanced sensing, and automated feeder reconfiguration.

For example, the smart grid can recognize irregularities within the utility grid and automatically adjust to increase energy efficiency and resiliency. Because of this, smart grids create environmental and cost benefits by reducing the amount of energy wasted and improving the efficiency of generation, delivery, and consumption.

Chapter 16

From Program Coding to Algorithms – Thinking Flow of Digital Work

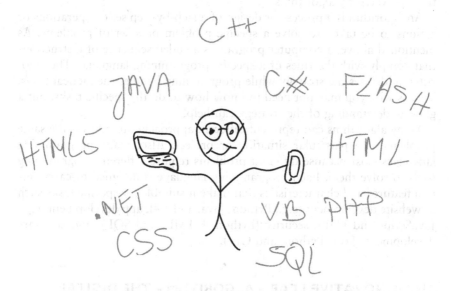

Computer coding and programming means to build up algorithms and instructions to run software applications via various programming languages.

Many people tend to ask, to accomplish digital transformation, does that mean we need to know how to code and learn computer programmings like Python or Java? The answer can vary. You need to know to code if you are designing or controlling an ICT system or device. On the other hand, if you are mostly using a system or device, then it is unnecessary for you to code anything. However, it does not hurt to gain some knowledge of coding and programming. This is like you drive a car but do not have to learn how to repair it, although it does not hurt to understand more about the vehicle, like how it works and the functionality of key parts under the hood, etc.

Knowing more about algorithms is a great approach to understand what coding and programming can do. An algorithm is more about an idea or a process or a design to solve a problem, while a program is more about

DOI: 10.1201/9781003305187-19

the technical procedures and executes one or more tasks by a computer. Algorithms are the building blocks for programming, allowing things such as computers, smartphones, and websites to function and make decisions. A program can execute one or more algorithms, or sometimes it may not need to use an algorithm where the app is simple.

We have briefly mentioned algorithms in Chapters 1 and 6 and now want to elaborate further on the magic functions they can play in the digital data world. Algorithms are the core of computing. In our current surroundings, it is not difficult to find several algorithms working to solve our daily life problems. Social media networks, Google search, ecommerce platforms, GPS applications, Netflix recommendation systems, etc., are all applications powered by algorithms.

An algorithm is a process or design of a step-by-step set of operations or actions to be taken to solve a specific problem or a set of problems. As mentioned above, a computer program is a coded sequence of instructions that comply with the rules of a specific programming language. The algorithm addresses the strategy while programming handles the tactical tasks. That is why you may not need to know how to do the specific tasks, but a good understanding of the strategy can help.

Many algorithms can represent different approaches to address the same problem for a particular situation or project. Also, many programming languages exist because different problems require different programming tools to solve them. Each programming language is designed to carry certain features and characteristics that make it suitable for specific tasks such as website development (e.g., Python, Java, and C#), app development (e.g., JavaScript and PHP), security (Python, HTML, and SQL), and software development (Java, Python, and C++).

THE INNOVATIVE LEAP – ALGORITHM – THE DIGITAL SOLUTION FLOW

An algorithm is not computer code; it is written in plain English and may be in the form of a flowchart with shapes and arrows, a numbered list, or pseudocode (a semi-programming language). Algorithms are all around us. An example of an algorithm is a cooking recipe, which tells us step-by-step instructions for preparing a dish or meal.

A very simple example of a computer algorithm would be that if you have a large unsorted list of numbers, it helps find the largest number from the list. If given a list of five different numbers, you would have this figured out in no time; no computer is needed. If you were issued a list of 5000 different numbers, clearly, you will need a computer to do this, and a computer needs an algorithm to guide the task.

We are in the age of algorithms, which get developed to handle our everyday tasks, and we cannot live or function without them. In the future, algorithms will work with AI to pioneer and predict our behaviors and digital performances. By nature, the algorithm provides the digital way of thinking of our time, paving the digital path or roadmap of problem-solving.

An effective algorithm is supposed to produce the correct outputs for any set of feasible inputs and execute efficiently with the fewest number of steps as necessary. A good algorithm should be designed so others can understand and modify it to specify solutions to additional problems.

Algorithms are involved in every part of computer science; nowadays, many decisions are made by algorithms. For instance, Google uses algorithms to optimize searches, predict what users will search, keywords and phrases, and more. In real business problem-solving, a big part of computer programming is about how to formulate an algorithm.

Below is a specific example of an algorithm converting binary (e.g., 1001) integers to decimal (e.g., 9) which is a primary guide on how a computer should work to process data:

Step 1	Multiply each corresponding digit of the binary number by the power of two and add it together	$1 \times 2^3 + 0 \times 2^2 + 0 \times 2^1 + 1 \times 2^0$
Step 2	Solve the powers	$1 \times 8 + 4 \times 0 + 2 \times 0 + 1 \times 1 = 8 + 0 + 0 + 1$
Step 3	Add up the numbers	$8 + 0 + 0 + 1 = 9$

NEW MINDSET FROM ALGORITHMS

Technical Outlook – The Characteristics of Algorithms

Computer algorithms work through input and output. As shown in the above example of binary converting to decimal, the algorithms take the input and then apply each step of the algorithm instructions to that result to generate an output. Such input leads to steps and questions that need handling in sequence. When each flowchart step is completed, the generated result is the output.

As mentioned already, one way to measure the efficiency of an algorithm is to count how many steps of operation it needs to go through to find the answer across different input sizes. A good algorithm should be both correct and efficient. Efficiency means that the solution will happen with as few steps – consuming as little computing power – as possible. Thanks to an algorithm, the user does not need to understand the problem to solve it.

A workable algorithm has four primary structures characteristics:

Finiteness	An algorithm should have a finite number of steps. It should end with the result after a finite time.
Input and output	An algorithm may have many or no inputs but should produce at least one output or result.
Definiteness	Each step must be well-defined, clear, and precise without ambiguity.
Effectiveness	Each step must be simple and productive.

Business Outlook – Algorithms for Better Business Results

Software applications powered by algorithms to leverage data can be used to optimize processes or create revenue streams. Algorithms are becoming thus a critical component of every business. Almost all of digital business insights and decisions will be data-driven and generated by effective algorithms.

Algorithms nowadays are no longer static but are evolving. Over the next few years, we can expect them to update and fix their approaches and behaviors automatically. We can also expect a growing trust we will feel toward AI. An AI algorithm is an extension of machine learning (ML) with the capability to instruct the computer how to learn to operate on its own. In turn, the device continues to gain knowledge to improve processes and run tasks more efficiently.

There is a caveat on Algorithmic decision-making which is expected to have the potential to eliminate, introduce or amplify biases or discrimination. This depends on how the software is deployed and the quality and representativeness of the underlying data used by the algorithm. For instance, when using ChatGPT, you may sometimes find its response is way off the mark or wrong. That might mostly be caused by our unclear and even misleading queries to ChatGPT in the first place.

Application Outlook – Algorithms Taking Over the World

Algorithms are used as specifications for data processing, mathematics, automated reasoning, and several other important chores, via data processing, calculation, analysis, and automated reasoning. Algorithms are becoming a digital part of our lives.

Algorithms allow software developers to produce efficient and error-free programs. Most importantly, there can be many different algorithms for the same problem, but some are much better than others. So that is the market competition where better algorisms win.

For example (refer to Chapter 8), SD-WAN as the next generation netwrok solution, currently has more than a dozen solution vendors in the market, each with their own version of SD-WAN. What would decide which SD-WAN solution can work out better? One major factor is it depends on the algorithm set that runs the SD-WAN solution. If the algorithm embedded is better, most likely the SD-WAN solution is better as well.

INSIGHTFUL PRACTICE

Algorithm-powered Sorting in Social Media

Algorithms in social media platforms function as the auto-recommender and can sort posts based on relevancy instead of publishing time. They would help prioritize which content a user sees first according to the likelihood from the users' historical preferences. For example, algorithms can recommend new posts to you based on what you scroll through in your Instagram feed or the stories posted by friends that appear first on the dashboard.

Algorithms are designed in a comprehensive way that takes into account different aspects of things. Some of these aspects are content based, and the algorithms seek to match a user's taste, based on the profile, to specific posts that the user may like. Whenever users show interest in a specific merchandise tag or category, they are also presented with similar items in the same category.

Algorithms can also operate collaboratively. Collaborative filtering allows matching users to other users who share similar interests. Along this way, users are directed to posts or videos they might want to see based on a user with a similar profile searching for that specific source.

Algorithms can become context aware, meaning they can individuate personal data like a user's exact geographic location and include it in the algorithmic calculations. For instance, the Facebook algorithm determines and recommends which posts people see every time they sign on and in what order those posts should show up. Essentially, the algorithm evaluates every position, scores posts, and then arranges them in non-chronological order of interest for each user.

As mentioned, underlying algorithm sometimes may malfunction, thus the effects of algorithmic applications can be positive and negative, depending on the quality and effectiveness of the algorithms in use. For instance, algorithms are created to increase awareness or interest in the digital society on a specific matter. Still, some users who are interested in sports may suddenly see in their feed an increase in posts concerning nutrition and diet, foreign cinema, or politics. Again such scenario is often caused by misleading or unclear queries or data gathered.

eCommerce Algorithms Driving the Business

eCommerce algorithms are used in several ways: to increase understanding of the customer and target customer, to learn from user behavior and on-site data, to predict customer preferences, and also to surface products for customers based on various factors. They benefit retailers by showing shoppers the products they are most likely to buy based on the data fed into the engine. For the shoppers, the improved relevance means a better shopping experience.

Algorithms for product recommendations are used extensively in eCommerce, which will help find the most suitable products for customers. These programs use a range of data to show relevant products more likely to be purchased. One popular instance is the A9 Algorithm which Amazon uses to decide how products are ranked in search results. The Amazon search engine system determines which products to rank within the users' search results.

On the other hand, the Google search algorithm is very famous for its success. It is a complex design that allows the Google search engine to find, assess, rank, and return the most relevant pages responding to a search query. As a matter of fact, the whole ranking system consists of multiple algorithms assessing various factors such as the page's quality, relevance, or usability.

Powered by deep ML algorithms, now we also have a new category of software called eCommerce Intelligence (EI) that enables merchandisers and marketers to maximize revenue from every visitor session. This way EI automatically personalizes and optimizes on-site shopping experiences for every visitor segment, targeting individual customer or group's shopping interest and needs more effectively.

Face-Recognition Algorithms with Controversies

A facial recognition system uses biometrics to map facial features from a photograph or video input digitally. Then it compares the data with a database of known faces to find a match. Facial recognition can help quickly verify a person's identity which however may raise privacy issues. Facial recognition uses computer-generated filters to transform face images into numerical expressions that can be compared to determine their similarity. "Deep learning" (Also refer to Chapter 6 and 17) usually generates these filters, which use artificial neural networks to process data.

Face detection and recogization are the two major tasks involved. The former focuses on facial landmarks and analyzes their spatial parameters and correlation to other features, while holistic methods view the human face as a whole unit.

Facial detection often uses Haar feature-based cascade which is mostly used to identify faces in an image or a real-time video.. This is a classifier, an effectual ML-based approach using a trained cascade of samples containing many positive and negative images. These include models for face detection, eye detection, upper body and lower body detection, license plate detection, etc.

As for face recognition, most modern facial recognition algorithms have some semblance of integrated deep learning and neural network, and this way they drive any facial detection and recognition system or software.

For instance, LBPH (Local Binary Pattern Histogram) as a face-recognition algorithm is used widely. It is famous because it can recognize a person's face from both front and side faces. In a controlled environment, LBPH is most likely to get great results.

Face Detection Software Apps

Chapter 17

From Managed to Self-Managed – Let Machine Learn to Act

This is a next-gen robot that is powered by artificial intelligence and machine learning.

Nowadays, we have heard a lot about the buzzword: machine learning (ML), which can pretty much speak for itself. In 1952, a then IBM employee and engineer Arthur Samuel coded the first computer learning program. The program was for playing the game of checkers, and the IBM computer improved the checker playing at the game: the more it played, gaining more moves made up winning strategies and incorporating those moves into its program, the stronger its capability of playing checker became.

Then ML takes nearly half a century to boom into real and wide applications. November 30, 2022 made a big day of breakthroughs for Artificial Intelligence (AI)/ML applications when ChatGPT (GPT stands for Generative

DOI: 10.1201/9781003305187-20

Pre-trained Transformer) was launched into the market. By January 2023, in just over 60 days, ChatGPT had taken the world by storm, setting a record for the fastest user growth when it reached 100 million active users.

ML is popular now because computation is abundant and widespread. Abundant and inexpensive computation has driven the explosion of data we are collecting and, therefore, the increased capability of ML methods. ML depends on big data and effective algorithms to happen.

Keep in mind that the key difference ML brings is that the machine starts to step into the human domain in doing things. We have been using a computer to collect and analyze digital data for decades. Our humans usually decide the final action taken according to the information abstracted from the data. For instance, digital data shows that five middle schools are pretty good in my area, then I need to decide which one I want my kids to attend.

But now, ML tries to make such decisions on our behalf based on data and information available. This is a revolutionary change. For another instance, to prevent a bank ATM from getting robbed at midnight, in the past, we would need someone in for such a red-eye duty to watch the situation. If some alarms came up, the person on duty needed to look and judge if the alarm was real or false and then quickly decide the next step, such as calling the police for help.

THE INNOVATIVE LEAP – MACHINES ARE TAKING OVER

Now we can have a computer-powered machine to do this job for us by monitoring, judging, and taking action, saving human resources and accelerating the time of proper response to reported problems. Someone may ask: is ML the same as artificial intelligence (AI)? The correct answer is that ML is an essential field of AI that covers a larger scope of technologies and applications.

Up-to-date, AI is very good at handling repetitive behavior patterns and forming a proper response accordingly. Like if in Amazon Book, the ML mechanism embedded in the portal finds a reader searching for books on history a few times in the portal, then it can start to proactively find all new publishing books on history and recommend them to this reader on time.

On the other hand, however, we find AI technologies not very good at detecting and simulating human emotions and feelings. For example, if we use a robot as a home assistant, it can handle many household duties fine, but if we feel happy and excited, the robot will not celebrate with us. The robot does not know how to calm us down if we feel sad or frustrated.

Anyway, that is a sophisticated area where AI still needs to improve big time (also refer to Chapter 6).

Luckily, ML, in most cases, focuses on AI strengths, namely, learning about repetitive behaviors, building out a pattern, and then making the right response to it. The amazing part of ML is that once it collects and analyzes sufficient data in certain application fields, it can function well and even beat human performance.

This is not a surprise, just like why we prefer to hire an experienced worker for a profession, because after so many years working within a certain profession, that worker becomes a professional who knows very well to handle all various situations and resolves many kinds of problems.

NEW MINDSET FROM MACHINE LEARNING

Technical Outlook – The Ways Machines Learn

A ML model is a program to be trained so as to recognize certain types of patterns. We train the machine a model over a set of data and provide it with an algorithm that can reason over and learn from those data. Hence ML can run predictive models that learn from existing data to forecast future behaviors, trends, and outcomes.

Today, ML algorithms are trained using three main methods called three types of ML: supervised, unsupervised, and reinforcement learning.

Supervised learning is the technique of ML to accomplish a task with training feedback. This includes training, input, and output patterns to the systems. Unsupervised learning, by its name, is a self-learning technique. It enables the system to discover the features of the input population on its own, and no prior set of categories is used.

Reinforcement learning is another ML training method through trial and error. A reinforcement learning agent is key; it can perceive and interpret its environment, take actions and learn based on rewarding desired behaviors and punishing undesired ones.

Deep learning is a trendy subset of a ML algorithm. By using multiple layers of neural networks, it performs in processing data and computations on a large amount of data. Capable of learning without human supervision, the deep learning algorithm can be used for both structured and unstructured types of data.

In deep learning, there is a famous convolutional neural network (CNN, or ConvNet) which is a class of deep neural networks. It is most commonly applied to analyzing visual imagery. A CNN can take in an input image, process the data, and assign importance (learnable weights and biases) to various aspects/objects in the image. In this way, the images can be differentiated from the others.

Like CNN, a recurrent neural network (RNN) is another class of artificial neural network. It connects nodes to form a graph along a temporal sequence. RNN is the state-of-the-art algorithm for sequential data, and its use cases include Apple's Siri and Google's voice search. It is the first algorithm that can remember its input using internal memory, which makes it perfectly suited for ML that involves sequential data.

Business Outlook – The Hybrid Intelligence

Here is a rule of thumb for ML: any repetitive task that creates many data will eventually be learned by computers. But experts agree that humans still tower over computers in general intelligence, creativity, and common-sense knowledge or understanding of the world.

The major difference is that humans use their brains to think and memorize, while AI machines depend on the availability and processed data. Humans can learn from past mistakes; intelligent ideas and attitudes lie at the basis of human intelligence. The optimal combination is the human intelligence supplemented by AI.

Hybrid Intelligence (HI), therefore, means the combination of human and machine intelligence. This might be the way to go and expand human intellect instead of replacing it. HI takes human expertise and intentionality into the mix; it makes reasonable decisions and performs appropriate actions, together with ethical, legal, and societal values.

Application Outlook – ML-powered Robots

ML can conduct the following three common tasks: Regression – predicting a continuous quantity of new observations using the knowledge gained from the previous data. The target variable is continuous. Classification: classifying the new words based on observed patterns from the earlier data. The target variable is discrete. Clustering is the process of grouping similar observations in one cluster and dissimilar comments in another.

Google's AlphaGo (also refer to Chapter 6) is an AI-powered robot that can now beat top human players at the game Go. How can that happen? Well, once the robot got programmed with thousands of smart moves recorded and collected by different top human players in all different scenarios, the robot becomes a super Go player. For each system and stage during the game, it has the "best moves in mind" ready to take the right steps.

In a nutshell, no matter how great his human counterpart is in the game, that is one person's brain. The AlphaGo robot, on the hand, is armed with smart moves from many top human players. Hence, it becomes one super smart head vs. many smart heads game. The robot now has the upper hand advantages. ML makes a vital tool for digital transformation in many applications by replacing or supplementing human work and duties.

INSIGHTFUL PRACTICE

Tesla's Auto Pilot

AI is at the core of Tesla's efforts to develop more advanced driver-assistance features for its cars and, eventually, toward fully autonomous vehicles. Tesla vehicles out of the factory since October 2016 include the hardware suite that may eventually enable full self-driving. This is an ML journey for Tesla as these cars drive around; they collect data from the cameras and other sensors. Then the Autopilot computers can use that data to learn by example.

Tesla's Autopilot relies on sensors to help drivers with steering, keeping a safe distance from nearby vehicles on the road, and other tasks. Built on a deep learning neural network and based on an unsupervised learning model of ML, here is how Autopilot works: it uses cameras and ultrasonic sensors to "know and see" the environment around the car. This suite also gives drivers awareness of their surroundings that a driver would not achieve naturally.

Tesla uses a custom AI chip called D1 for training the machine-learning algorithm behind its Autopilot self-driving system over a dedicated computer, Dojo, which runs a neural net for vision, sonar, and radar processing systems. Tesla's large fleet also provides other advantages in computer vision, behavior prediction, and decision-making. In addition, Tesla uses deep neural network algorithms to train its autopilot with collected real-world data. Via AI, or ML techniques designed, Tesla helps machines think more like human drivers, supporting its advanced driver-assistance system, Autopilot.

The Storm of ChatGPT

What is ChatGPT? Let's see how ChatGPT introduces itself during an interview by Freethink.com (source: https://www.freethink.com/robots-ai/ai-chatbot-chatgpt):

> I am an AI assistant trained to assist with a variety of tasks, including answering questions, providing information, and helping with decision making. I am not a physical being, but rather a program that runs on a computer. My creators at OpenAI have trained me to understand and generate human-like text, so I can interact with people in a natural way. Is there something specific you would like to know?

In this same interview, ChatGPT also claims,

> "as a large language model, I have been trained on a vast amount of text data and can generate human-like text on a wide range of subjects. However, I am not an expert on any specific topic and do not have specialized knowledge beyond what I have learned from the text data I was trained on. In other words, my responses are based on the patterns

ChatGPT can generate new content for users.

and information I have observed in the text data I have been trained on, but I do not have any additional knowledge or expertise beyond that."

In its interactions with humans, ChatGPT works by attempting to understand user's prompt and then generating strings of words that it predicts will best answer the user's question, based on the data it was trained on. This training model is called Generative Pre-Trained Transformer (GPT) which is a language model that relies on deep learning to generate human-like texts based on a given prompt. A user "feeds" the model with a sentence query, and the transformer creates coherent paragraph-based information extracted from the public Internet datasets.

The biggest use case of ChatGPT is content creation. Converting words into compelling stories can keep the reader engaged and wanting to know more about your business brand. ChatGPT can do this by changing the tone of the words and making them persuasive in seconds. With regards of blog writing and posting, despite the blogs' benefits, writing these long blog posts can be overwhelming. Not anymore. With the assistance of ChatGPT, blogs can come up quickly.

Concerns always accompany each technical breakthrough. In this case, will chatGPT take away human jobs? The answer is not really. Although it can be used in industries such as customer service, marketing, and content creation, by nature, ChatGPT is an advanced tool and it will not take away real jobs. It will assist in improving existing jobs and automating certain tasks. In a similar analogy, has email replaced post mail? We know the answer is yes to a certain extent, but not completely.

Starbucks' Deep Brew

Is Starbucks – the world-renowned coffee shop, becoming a technology company as well? Yes, amazingly, it is. Today, Starbucks is a tech company. Technology at Starbucks encourages big ideas and forward-thinking solutions. Other businesses may boast better coffee, tastier food, or a more modern ambiance, but none can currently compete with Starbucks' app, AI engine, or financial features.

With their technology initiative called Deep Brew, Starbucks is starting and working on a broad suite of AI tools, hoping they can help elevate every aspect of the business and the in-store and customer experience. The initiative allows the coffee giant to innovate and apply ML and AI to personalize its drive-thru experience and automate time-consuming tasks such as inventory management and prevention.

Starbucks has been using reinforcement learning technology that allows a system to learn and make decisions in complex, unpredictable service environments based on customer feedback. The approach helps personalize customer services for those who use the Starbucks® mobile app. The ML algorithms developed by Starbucks also take in data like the weather, popularity, store inventory, time of day, and community preferences. To make all the data understandable to the machines, a lot of entity and text annotation will need to be done to train the system.

Via reinforcement learning technology, customers' previous orders and preferences are analyzed. Then tailored new recommendations are made to them. This way, customers can find new flavors they enjoy and get a more intimate and personalized experience. As a result, Starbucks gets more coffee sold.

Chapter 18

From Social Media Toward MetaAlverse – Our Next Digital Space

Social media examples include Twitter, email, Meta (Facebook), LinkedIn, etc.

Like the still brewing Web 3.0 concept (refer to Chapter 10), we often must experiment with technologies and their applications. Many such experiments later may prove successful, but many others will be just bubbles and duds. But we cannot stop trying and moving on. In late 2021, social media giant Facebook announced that they officially changed their company name to Meta with the mission to build a metaverse – the virtual space for everyone.

While social media helps people to communicate and connect, the metaverse might join what are now separate apps from each company in the future. According to Facebook, for example, one might be able to mix and match virtual glasses, makeup, and apparel from any seller and bring them to games, concerts, meetings, or other metaverse events.

DOI: 10.1201/9781003305187-21

Facebook's vision is that the metaverse refers to much broader and highly immersive online experiences. Examples are virtual reality (VR), augmented reality (AR), interactive video, cryptocurrency, and Web 3.0. Metaverse is essentially targeting to take the online experience to the next level, creating a new digital world where people can socialize, conduct business, and experience immersive entertainment.

By early 2023, however, the successful launch of AI breakthrough-ChatGPT (refer to Chapter 17) has dwindled the metaverse idea and momentum. Facebook (Meta) decided to switch business focus. While Metaverse sounded like a big idea only internally to Meta, ChatGPT has just got its global market success and proof. Meta now plans to start with generative AI tools as well and then focus on developing AI personas that can assist people in a variety of ways.

ChatGPT has completely changed Internet search. Now, we can get human-like responses to queries. AI's uses for consumers and businesses are indisputable: for companies, repetitive and boring tasks can now be carried out efficiently using chatbots. By the end, who qualifies and decides the real market-breakthrough products? The market. In the long run, AR and VR are still hot applications of high technologies, but the best way to bring them to market turns out to be via AI.

THE INNOVATIVE LEAP – AR GOING BEYOND VR

Technology-wise, VR and AR are real deals already being used on many occasions. In 1968 at Harvard, computer scientist Ivan Sutherland who is named the "father of computer graphics," created an AR head-mounted display system. This was the first AR technology developed.

VR is a computer-generated environment, simulating scenes and objects that appear natural, giving the user the feeling that they are immersed in their surroundings, acting as a real presence in places in the physical world or imagined worlds. This environment is viewed via a device known as a VR headset or helmet. Specifically, VR simulates the vision of a 3D environment where a user appears immersed while browsing or experiencing it. The 3D environment is then controlled in all 3D by the user experiencing it.

What is the primary difference between AR and VR? Both technologies can replace and add digital elements to visual perception. While VR immerses the viewers in a virtual world, AR goes to the next level by adding digital elements to physical reality. Hence, AR is an enhanced version of the real physical world. It is achieved through digital visual factors, sound, or other sensory stimuli delivered via software technology. AR enables us to interact with virtual things as if in a real scenario via the device supporting the AR features with the lens to scan the object where the digital impression is to be generated.

NEW MINDSET FROM VR AND AR

Technical Outlook – Real and Virtual Integration

VR uses computer technology to generate simulated environments and a three-dimensional user experience. Traditionally, viewers have a screen in front of them, but now users get immersed in and interact with 3D worlds via the helmet.

Immersion and interaction are most important for VR to provide a surreal and enjoyable experience for the user. The VR and AR-powered metaverse is essentially a merging virtual, augmented, and physical reality and blurs the line between one's interactions online and in real-life.

AR technology combines virtual information with reality. Its technical means include multimedia, 3D-Modeling, real-time tracking and registration, intelligent interaction, sensing, and more. Some of the challenges VR faces include having a limited user experience and a limited number of use cases. Digital fatigue and the high price of developing new technology are some other cons.

Business Outlook – The Combined Product Design Power from AR and Generative AI

The advent of augmented reality (AR) and generative artificial intelligence (AI) can make an innovative game-changer for physical product design and are a dynamic duo when it comes to creating designs that are sustainable and truly tailored to the customer's needs and environment.

Using building or construction design as an example, one of the most exciting applications of AR in design is its ability to help the designer analyze sites to come up with the best arrangement for energy efficiency. Using AR the designer can visualize how a building's orientation affects solar exposure and shading patterns. This helps designers decide about window positions, shading utilities, and other elements that impact energy efficiency.

Generative AI, on the other hand, helps designers in generating designs that are specific to the needs of the project. AI algorithms can generate multiple design options, providing designers with a broad range of options, each with different features that can be refined and developed further.

AR is also revolutionizing how we communicate and present designs beyond traditional static 2D drawings or 3D renderings. All parties involved can experience the design onsite through AR, giving them a much better understanding of the design and its context. Together, AR and generative AI bring a new level of responsiveness to design. Instead of a generic structure, designers can now create products that are unique and truly in tune with the right environment.

Application Outlook – More Immersive Customer Experiences

The future AR/VR/AI devices will be more personalized and accessible and provide well-designed experiences. The metaAIverse roadmap prospect four future types: AR, lifelogging, mirror world, and VR/AI. An example of the application of AR in medical education would be an AR T-shirt as an anatomy lab that allows students to study and examine the inside of the human body with AI enabled reponses to queries.

The metaAIverse enables marketers to inject their brands into that experience, making each user the show's star. Instead of seeing famous actors, models, or rap artists hawking the latest styles, clothing, or fashion brand in the metaverse, it can allow each of us to try on their latest looks with clear illustrations and vivid descriptions.

For the hospitality vertical as another example, customers can experience seamless booking processes to one-of-a-kind stay experiences, and the service will be more optimized with AI enabled data and choices. With the metaAIverse, customers can take a 3D virtual hotel tour, receive top recommendations, and research the hotel's location where they are considering staying using VR/AR/AI platforms and hardware.

INSIGHTFUL PRACTICE

NASA and VR

NASA – the US space agency's interest in VR dates to the early 1990s, The VR lab, founded in the early 1990s, was involved in training astronauts during the Space Shuttle era.

According to NASA, VR/AR tech is becoming an increasingly strong staple in life aboard the International Space Station (ISS), often used to help astronauts to complete tasks or activities in new or more accessible ways. Astronauts typically spend hours training in simulators for every minute they fly in space. This way, they are familiar with planned activities and can react quickly to unusual events. They rehearse their own duties once and again and learn the teamwork vital to successfully overcoming a hurdle.

The simulations of social scenarios can mitigate stress and establish connectedness in an isolated and confined environment (ICE). VR simulates for astronauts before leaving earth, the environments in space, such as the ISS. For example, Space Explorers: the ISS experience was filmed aboard the ISS over two years, and used to simulate one of the best VR space environment for astronauts training on the ground.

VR and AR also enable scientists to run missions impossible and visualize environments that are typically difficult or even impossible to visit, such as the surface of Mars. Such immersive visualization begins with a high-quality reconstruction of the environment under study. VR technologies developed under Goddard and NASA research and development programs make designing spacecraft, instruments, and repair missions easier, allowing engineers to experience space before building it.

The Jurassic VR World

VR technology has opened up new opportunities to make films that take place in virtual realities via viewers wearing VR headsets. In 2016, the Cannes Film festival, for the first time, started to screen short VR films and presentations, and the festival had an area exclusively dedicated to VR.

This technology has started to provide a different type of viewing experience. It provides viewers with more freedom in what they are watching. However, whether the technology offers the same storytelling or quality as the film remains debatable. For instance, director Steven Spielberg describes it as dangerous "because it distracts the viewer many latitudes from the direction of the storytellers." In a nutshell, the viewers now have many distractions to cope with during their film watch.

It may take a while until moviegoers all have the opportunity to experience VR movies through the lens of VR headsets. However, this technology has inspired incredible feats in film production that viewers are currently witnessing, including StageCraft with its soundstage, commonly referred to as the Volume, which is an on-set virtual production visual effects technology composed of a video wall designed by Industrial Light & Magic (ILM), as a revolutionary visual effects technology.

In the post Covid-19 era, with many movie theaters closed, and many movie lovers wary about returning to public spaces, this could be an opportunity for VR platforms like the Oculus to grab cinephiles who are desperate for more immersive experiences than they can get sitting on their couch watching a 40-inch screen. *Star Wars* already has a few impressive VR games for the platform, including *Vader Immortal* and *Tales from the Galaxy's Edge*. Now *Jurassic World* is getting into the virtual reality VR business with a new title called *Jurassic World: Aftermath*.

Virtual Field Trips for School Education

A virtual field trip is a guided venture with online digital experiences. It organizes a collection of pre-screened, thematically based websites into a structured online learning experience. VR and AI make vital tools for education because they enhance learning and memory recall. Immersing in virtual locations helps students organize and remember information spatially. This is because they associate the AI generated information with visual features in the environment.

Virtual field trips began to showcase in the mid-1990s and have become increasingly popular. The biggest pro is that they can complement traditional classroom settings. Furthermore, they are also a great tool for parents and homeschooling groups, specifically at the elementary and middle school levels.

Field trips are enriched learning experiences that get students out into the world, exploring reality and students' interests, enabling students to discover new things and learn in real environments. enrich and expand the classroom curriculum. In education, learning through experience has been claimed to be the most effective learning approach, or one image speaks thousands of words. Studies have shown that field study increases the quality of learning and retention by 70–90%. A field trip memory may stay and influence a person in their lifetime.

Today AR/AI powered field trips strengthen children's observation skills by immersing them in sensory activities, and increase their knowledge in a particular subject area and raise awareness of their community. Students are encouraged to think critically and apply the concepts learned in class to the real world.

Hence the AR/AI advantages are clear: virtual field trips are not constrained by distance and are more cost-effective than traditional in-person field trips. They also eliminate the need for transportation, decrease lost instruction time spent on travel, and involve fewer safety concerns, e.g., no parental permission is required.

This person is wearing a helmet to enjoy the VR and AR experiences.

Chapter 19

From 4G LTE to 5G – Automations Enabler

This illustration shows the wireless technology evolution journey from 3G to 5G.

What is the biggest buzzword of network technology in the 3rd decade of 21st century? Some people say it is 5G, and G here refers to generation. But 5G is not just a buzzword; it will surely change the world. Many of us might have taken technological progress for granted, and it seems to come up naturally. But keep in mind technology progress may come at different paces and scales. Sometimes it is evolutionary with progress, and sometimes it is revolutionary with breakthroughs.

Take wireless technology as an example; from 1G to 2G; it was all about voice mobile phone services from the beginning to a standard

DOI: 10.1201/9781003305187-22

telecommunication service; then from 2G to 3G, it went to a little higher speed with mobile voice service maturity; from 3G to 4G LTE, with the emerging smartphone, the wireless focus started to shift from voice to data communications, up to 5–12Mbps speed, peaking at 50 Mbps.

From LTE to 5G, the technology goes beyond regular phone services to the next level of digital application enablement, from advanced IoT and edge cloud computing to remote automation. By design, 5G can support a 100× increase in traffic capacity and network efficiency. The average 5G speed is 100 Mbps, up to 20 Gbps.

The greater capacity offered by 5G will allow networks to support more devices and enable more data-intensive tasks. What are the trademarks of 5G services? To consumers, 5G will bring three primary benefits : faster data speeds, lower latency, and increased connectivity. But 5G is more for advanced IoT and apps like driverless vehicles and together with AI/ML makes the enabler of industrial automation.

THE INNOVATIVE LEAP – 5G FOR AUGMENTED IOT

So, we can tell that 5G compared to 3G and 4G leaps in wireless technologies and will lead to a leap in service applications as well, such as real event streaming, telemedicine, driverless vehicles, robotics, and virtual reality gaming. 5G is playing a pivotal role in the ongoing 4th Industrial Revolution.

The question many people have in mind is how 5G achieves this speed and bandwidth leap. The answer is due to three major factors: increased spectrum, multiple input, multiple output (MIMO) antennas, and all fiber optic transport.

Increased spectrum and greater capacity means additional mobile spectrum above 6 GHz, including the 26–28 GHz bands or millimeter (mm) Wave, which provide significantly more capacity than current LTE and 3G mobile technologies. The mm spectrum and greater capacity will enable and serve more users, more data, and faster connections.

5G will use MIMO antennas with very large numbers of antenna elements or connections. MIMO can send and receive more data simultaneously. The benefit is that more people can simultaneously connect to the network and maintain high throughput.

For 5G, the fiber optic-powered "core network" is now being redesigned to better integrate with the Internet and cloud-based services. The network also includes distributed servers or edge cloud across the network, improving response times (reducing latency). As a result, we can expect 5G will cut latency dramatically in mobile applications.

For instance, the average latency from 4G LTE is 20–30 ms, while 5G can go down to 4–5 ms, making real-time and instance, high bandwidth data like a real sports event coverage possible. For driverless vehicles, ultra-low latency in 5G can make the latency as short as 1 ms, which is real-time literally.

NEW MINDSET FROM 5G

Technical Outlook – mmWave and Network Slicing

The proposed 5G architecture consists of four layers: the network layer, controller layer, management and orchestration layer, and service layer.

Mid-band between 1 and 6 GHz can facilitate speeds up to 1 Gbps. Some major carriers' wide-reaching 5G networks operate in the mid-band. But to reach the top speeds associated with 5G, carriers also need mmWave spectrum, which takes advantage of the very high-end 6GH> of the wireless spectrum.

Plus, for a unified, more capable 5G wireless air interface, 5G New Radio (NR) is becoming the global standard. 5GNR is expected to deliver much faster and more responsive mobile broadband experiences. It will also extend mobile technology to connect and enable a multitude of new industry apps.

Advanced features of 5G, including network function virtualization and network slicing for different applications and services, will be managed in the core. Network slicing is a key feature of creating multiple unique logical and virtualized networks over a common multi-domain infrastructure and enables a smart way to segment the network for a particular industry, business, or application.

The 5G core uses a cloud-aligned service-based architecture (SBA) to support authentication, security, session management, and traffic aggregation from connected devices, all of which require the complex interconnection of network and cloud functions.

Business Outlook – The Edge Automation Enabler

What is 5G about, and where can they be best used? Two key directions to remember: First, faster and higher speed. While 4G usually can go to 50–100 Mbps in speed, 5G will go 100 times faster to 1–30 Gbps level using different spectrums. Second, wider and more robust applications: The 5G high speed means faster downloads for apps and support for more connected devices than ever before. For example, we must rely on 5G technology to control driverless cars.

5G wireless technology delivers ultra-low latency, more reliability, higher multi-Gbps peak data speeds, massive network capacity, and increased availability. Together this will deliver a super user experience to more users, in other words, higher performance and improved efficiency that will empower new user experiences and connect new industries.

As already mentioned, the beauty of 5G network slicing is to allow multi-application traffic to transit through the same platform with security and specific policies and SLAs. For example, emergency services could operate on a network slice independently from other users.

Application Outlook – New Horizon for Low Latency IoT

Specifically, the rollout of 5G will provide advances in three major areas, or the so-called "5G triangle": the first one is called uRLLC: Ultra Reliable Low Latency Communication use cases like live streaming. Then mMTC: Massive Machine Type Communication use cases for IoT. Lastly, eMBB: Enhanced Mobile Broadband – high-speed use cases like fixed wireless for home and business Internet services.

5G and edge computing (also refer to Chapter 23) will enable AI at the edge with high-performance edge cloud and IoT. The adoption of 5G networks is bringing new opportunities to various industries. A key aspect of 5G is that it enables software-defined networks (SDN) for easier data and device management on intelligent software layers. When integrated, 5G and edge cloud will help significantly improve the operation of IoT applications by delivering: ultra-low.

Will 5G replace Wi-Fi? The two technologies will likely coexist, as Wi-Fi uses a public network, while 5G can go private. Organizations will make strategic decisions about how their IT infrastructure should evolve as network rollouts progress. In some cases, 5G can help address many pain points associated with Wi-Fi deployments.

INSIGHTFUL PRACTICE

Onsite Automation – 5G and AI Synergy

While old wireless technologies were for basic voice and data services, 5G was born in the cloud era and explicitly designed for IoT apps since 5G can connect billions of sensors, such as video cameras and bank ATMs, to edge data centers for AI processing.

5G and AI are two essential ingredients that fuel future innovations, and they are inherently synergistic – AI advancements can help improve 5G system performance and efficiency. In contrast, the proliferation of 5G-connected devices can drive distributed intelligence with continued enhancements in AI learning and inference.

For instance, AI powered robots can be controlled over vast distances using 5G technology. Robots' operations must interact with their environment in real-time and process huge amounts of information via instant communications. It is the optimal choice with 5G's lower latency and higher bandwidth than legacy wireless connectivity.

5G and AI infrastructures are not efficient because they are deployed and managed separately today. The trend is for enterprises to run AI and 5G on the same computing platform, reducing equipment, power, and space costs. Such synergy also provides greater security for AI apps due to 5G slicing capabilities. For telecommunication carriers, deploying AI apps over 5G opens up new industry use cases and revenue streams. Every 5G base station can be converted to an edge data center to support 5G workloads and AI services.

5G will speed up the integration of other technologies, while AI will allow machines and systems to function with intelligence close to the levels of humans. In a nutshell, 5G speeds up the services on the cloud while AI analyzes and learns from the same data faster.

Autonomous Driving and 5G

5G autonomous vehicles are a widely adopted use case of uRLLC. With up to 20 Gbps peak data rates, 5G will enable real-time video and audio entertainment in autonomous cars. But more importantly, 5G's fast and reliable data connection will allow downloading a sophisticated 3D map in near real-time.

Since real-time autonomous decisions need to be taken within a fraction of seconds, the low latency feature of 5G enables cars to receive information at turbo speed and take prompt actions that can avoid them from hitting obstacles on the way, hence the inclusion of 5G technology in the automotive industry will unlock opportunities for connected cars, connected bus-stops, smart metros, and autonomous vehicles. Such facilities will translate into ecosystem scalability, navigation and augmented reality, fleet monitoring, safety, and sustainability.

Self-driving cars use hundreds of sensors to make vehicles faster and smarter. These sensors monitor their surroundings and make informed decisions. Self-driving vehicles must be connected to the high-speed Internet, even if edge computing hardware can solve small computing tasks locally. No current self-drive car needs a continuous connection to the Internet to drive, but it needs intermittent connectivity. The computation for self-driving is all on the vehicle.

Self-driving cars generate and keep a map of their surroundings based on various sensors embedded all over the vehicle. Radar sensors, for example, monitor the position of nearby vehicles. Video cameras detect traffic lights, read road signs, track other cars, and look for pedestrians.

5G will effectively allow the necessary transport of data and intelligence to empower the automated vehicle. Autonomous cars cannot even handle simple things, such as detecting a pedestrian or swerving around a curb, without fast data processing capabilities and actions as fast as human reflexes.

Augmented IoT and 5G

5G is becoming the new standard for cellular networks. The Internet of Things (IoT) is rapidly developing and expanding. 5G will increase cellular bandwidth by huge amounts, making it much easier for the IoT to connect large numbers of devices.

Chapter 15 describes IoT as a system of interrelated computing devices, objects with sensors, and mechanical and digital machines. IoT also serves pets or people who are provided with unique identifiers (UIDs). IoT devices can automatically send data over a network without requiring human and computer interactions.

Basic IoT, e.g., smart home apps, can live with lower speed and higher latency but augmented IoT requires real-time two-way communications, high speed, and low latency to function effectively. Since 5G operates ten times faster than current LTE services, this increase in the speed of IoT devices in communication, sharing data, and functioning effectively is better than ever. Besides smart home devices, nearly all IoT devices will benefit from greater speeds, including those with healthcare and industry applications.

Because of 5G and even 6G in the future, IoT has the potential to be limitless. Advances to the industrial IoT will be accelerated through increased network agility and integrated AI. The capacity to deploy, automate, orchestrate, and secure diverse use cases will increase at hyper-scale. For instance, manufacturing builds smart factories with AI, AR, and robotics.

Other augemented IoT and 5G use cases include: public safety requires real-time video, secure communications, and media sharing. Advanced supply chain management via IoT sensors increases production, streamlines processes, and reduces costs. Media via 5G introduces new capabilities such as high-definition streaming and virtual reality.

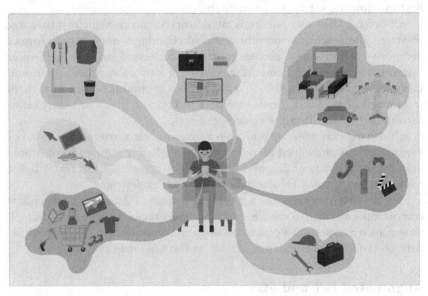

This illustration shows the broad fields for 5G applications from eCommerce and autonomous driving to communications.

Layer 4 – Digital Solution and Application Models – Operation Transformation, Customer Experiences, and ICT Ecosystem

Digital technologies are not about some point solutions addressing the individual and fragmented issues only, and the real strengths of digital solutions are from the unified platform, integrated fabric, and converged ecosystem. When thinking and planning for digital transformation roadmaps, we must envision the paradigm shift, new business models, and architectural innovation. We cannot just think about how to do better on top of the existing models, and that is far from sufficient in unleashing the potential of the digital era. Layer 4 of the Cognitive Model of Digital Transformation advocates that only the sky is the limit in this digital transformation process.

> The number one benefit of information technology is that it empowers people to do what they want. It lets people be creative. It lets people be productive. It lets people learn things they didn't think they could learn before, and so in a sense, it is all about potential.
>
> – Steve Ballmer, Former CEO of Microsoft Corp.

> Most of the executives I talk to are still very much focused on digital largely as a way to do "more of the same," just more efficiently, quickly, and cost-effectively. But I don't see a lot of evidence of fundamentally stepping back and rethinking, at a basic level, "What business are we really in?
>
> – John Hagel III, Co-Chairman at Deloitte LLP Center for the Edge leaders

DOI: 10.1201/9781003305187-23

Layer 4 – Digital Solution and Application Models – Operation Transformation, Customer Experiences and ICT Ecosystem

Chapter 20

From Point Solution to Digital Platform – The Single Pane of Glass

A digital platform converges technologies and solutions together to provide a one-stop service to the end users or consumers.

A centralized or called single pane of glass digital platform is the user or customer interface for modern ICT solutions and services, and it can be either internal user or customer facing. You may already use quite a few such digital platforms as speak; for instance, are you buying stuff online, like from Amazon.com? If yes, then you are utilizing Amazon's online platform. Do you use smartphones like iPhone or Samsung? If yes, you are using iPhone and Samsung's smartphone platforms. Do you use social media such as LinkedIn, Twitter, or Facebook? If yes, you are already on their digital platforms.

Digital platforms also make new business models that drive internal business structure changes across traditional organizational structures, silos, policies, and technology investments. A new digital platform often brings about a different organization and talent model. It also makes changes in mindset, policies, and processes.

The modern ICT infrastructure comprises hardware like end-user devices, servers, fiber optics, silicon chips, sensors, firewalls, and other

DOI: 10.1201/9781003305187-24

electronics. Physical sites of the infrastructure would include data centers, wireless signal towers, cable landing stations, control centers, branch offices, headquarters, consumer homes, etc. Practice before was that we used different systems and modules to handle networking, communication, storage, security, analytics, etc., separately or in separate silos.

To make all these elements in ICT solutions work seamlessly and concurrently together, now we need a centralized digital platform or a center, most likely dwelling in the cloud nowadays, which enables data-driven solutions and handles an end-to-end business process necessary to achieve the enhanced experience for customers, employees, and partners. In other words, these elements must be aligned and integrated to create better experiences for customers and users.

Multiple components, including hardware and software, make a modern digital platform possible, typically a data-ingestion engine, a network control center, a security management center, and an AI/ML transactional drive to perform tasks or rules-based activities. In addition, a digital platform is also powered by an analytical engine and, increasingly, APIs, virtual tunnels, and tools monitoring regulatory compliance.

THE INNOVATIVE LEAP – A SINGLE PANE OF GLASS

Today, we cannot function effectively in the digital age without engaging a digital platform, either as a business or a consumer end user. A digital platform has two major and attractive traits: first, it combines and consolidates different business functions powered by technologies – namely, an all-in-one approach to creating first-class customer experiences. For instance, if you can only view merchandise in an eCommerce web portal, but it does not allow you to check out and buy things at the same site, you will not like such a portal, will you?

Second, a digital platform should be part of an overall digital business ecosystem. This means a network platform allows the users to set connectivity between different branch offices. But what about the users who also need to set up their cloud functions in AWS, Azure, or Google Cloud? The platform should supply APIs to link to those cloud providers. What about the users who also want to set up more security features? The platform should have APIs linking to professional security vendors as well.

Why do smartphones are more appealing to users than legacy cell phones? Because the old cell phones only provided the so-called "point solutions" like making a phone call, or checking business emails, with minimal online functions. A smartphone, however, comes in via an all-in-one platform and works like a platform to control multiple key functions. If you need something specific, it allows you to use and link to 3rd party applications to handle it, e.g., you can set up smart home apps that automatically control indoor temperature via Google Nest. That is why we call smartphone a platform-based service,

Now you may find your car becoming a digital platform too. A car used to be a transport vehicle from point A to B. Now you can use your cars to check locations, road conditions, direction navigation, run trouble diagnosis and maintenance status, play audio, video, and games, and make hands-free phone calls while driving.

NEW MINDSET FROM DIGITAL PLATFORM

Technical Outlook – Platform-based ICT Solutions

The major technology advantage of a digital platform is that integrated solutions are more effective and efficient than point solutions. An integrated solution makes a full service happen, while a point solution only makes a single function possible.

An application platform is a framework of standard operations for application programs. From a technical operation standpoint, an application platform operates across five principal areas: development tools, execution services, data services, networking, operating systems (OSes), and cloud services.

Regarding business technology, a digital platform can be regarded as the gateway or center for transactions of information, goods, or services between producers and consumers. The platform can also interact with the ecosystem of platform operators, developers, and regulators.

Business Outlook – Digital Platform Thinking

A digital platform is not a product in itself. Instead, it is an established interface erected on present-day cloud technology that makes the evolution of software or programs easy. The services, applications, and solutions on the platform are the products that you, as a customer, will interact with and pay for. Digital platforms deliver new business capabilities in innovative and transformative ways and create value by enabling interactions between consumers and producers.

Digital platform thinking is revolutionary about how markets work in the digital era, such as how consumers, companies or brands, competitors, and others interact and create value. Digital platforms bring value by gathering, synergizing, and connecting customers, producers, and providers (ecosystem partners). The platforms will automate interactions and transactions in a multi-sided model with a network effect.

The digital platform experiences shape customer expectations with digitally native organizations. Such a platform drives all organizations to adapt their business models to become more convenient, responsive, and personalized. Successful platforms enable exchanges by reducing transaction costs and enabling client-facing innovation. With the advent of the Internet and 5G, these ecosystems facilitate digital platforms to scale in robust ways that traditional businesses cannot.

Application Outlook – One-stop Platform Services

Digital platforms ease development, reduce capital and maintenance costs via a service model. By enabling information and communication flows on the Internet, as well as digital trade across the globe, digital platforms assure heightened interoperability, helping accelerate digital transformation with AI and built-in accelerators.

Such "new engines" increase consumer choice and convenience, improving industries' efficiency and competitiveness. They can enhance civil participation in society as well. Famous digital platforms include Google Search, Facebook, Amazon Web Services, Amazon Marketplace, Android, Uber, Airbnb, Waze, WeWork, Twilio, and even Bitcoin.

To the clients and consumers, a digital platform provides one-stop shopping or the whole life cycle of a service. It also immerses with the overall digital ecosystem of services and brings apps to the next level, which is exactly what digital transformation is about. We use digital platforms for entertainment, news, transportation, accommodation, finding jobs and employers, downloading apps, and for many other purposes.

INSIGHTFUL PRACTICE

Salesforce.com – A Renowned CRM Platform

Salesforce.com is a renowned customer relationship management (CRM) platform. To many firms, it is a dream coming true that they finally have a single portal that can handle CRM nationwide or even globally. It helps marketing, sales, commerce, service, and IT teams work as one from anywhere – so companies can keep their customers happy everywhere.

Salesforce.com functions as the central point to consolidate, trace and manage sales activities, and it is the app development platform that extends CRM's reach and functionality. Companies use Salesforce.com to understand their customers, connect with them on various levels and help grow their customer base. Fitting into a diverse selection of clouds and applications, Salesforce.com is also used by companies to assist with marketing, tracking sales and spending, and analyzing performance.

A cloud-based SaaS company, Salesforce.com hosts an application that customers can access online. It operates on a pay-as-you-go subscription-based business model. Since all data and information are stored in the cloud, Salesforce.com can be accessed from any device at any time.

One of the most significant Salesforce.com advantages is a single source of data storage. Those days are long gone when we had to painstakingly acquire data from a million other sources. Salesforce is hitting the importance of generating comprehensible and convenient solutions that streamline how teams work together.

One of the reasons that Salesforce.com is so popular is that it carries features like no other CRM software; components, such as contact management, workflow creation, task management, opportunity tracking, collaboration tools, customer engagement tools, analytics, and an intuitive, mobile-ready dashboard. Today more than 150,000 companies use Salesforce.com tools in different departments and business areas.

Taobao – China's No.1 eCommerce Portal

For mostly political and cultural reasons, major Western eCommerce portals like Amazon.com have not fared well in China. If you plan on living in China and happening to know or learning Chinese there, using Taobao and Tmall, the country's two biggest online shopping sites, can be a good choice. There is no English version of these portals, so you need to be able to read Chinese to use them or use online translators to get over language barriers.

Taobao is a leading online shopping web portal because of the scale of the market and its users there. The demand for Taobao is rapidly increasing as time goes on, as more and more Chinese users are getting online and attracted to the network. About 1.1 billion people in China were actively shopping on their mobile devices in December 2022. Taobao was one of the most popular shopping apps in the country, with over 876 million monthly active users as of that month (source: https://www.statista.com/statistics/1327377/china-taobao-monthly-active-users/#:~:text=About%201.1%20billion%20people%20in,users%20as%20of%20that%20month).

This presents strong network effects in China's huge population base, which increase the value of Toabao's brand with over 1 billion product listings in this online marketplace. Taobao is owned and operated by Alibaba, now the world's largest mobile commerce company. In terms of scale, Alibaba is way bigger than Amazon.

Taobao marketplace operates online via B2C (business-to-consumer) and C2C (consumer-to-consumer) strategies. This model allows retailers to market their products online utilizing social shopping and has experienced significant growth in popularity and success. As of early 2023, Taobao has far over 875 million monthly active users, more than 8 million stores, and over 1 billion product listings, and accounts for nearly 60% of the total e-commerce sales on the China mainland, (source: https://marketingtoc hina.com/full-guide-to-selling-on-taobao-foreigner-merchants-or-small-busi- nesses/#:~:text=It%20has%20far%20over%20875,commerce%20sales %20on%20the%20mainland).

This shows an example of customer using an eCommerce portal.

Considering how many transactions Taobao handles annually, the plat- form does not charge sellers or buyers a transaction fee. Instead, it makes money by offering SEO (search engine optimization) – like advertising to sellers in the same way that Google does.

Zoom – Boosting Unified Communications (UC)

Unified Communications, or Unified Communications as a service (UC or UCaaS), is a cloud-based platform. UC or UCaaS bundles several core business communication and collaboration functions into one unified plat- form and interface.

The three basic elements – presence, user interface, and data integration with voice capabilities – are the required components of a UC solution, but the other components provide a more full-featured solution.

As COVID-19 broke out in early 2020, remote work or work-from-home became a new norm, and so did the UC tools for work-from-home. One of the biggest successes is Zoom, which has developed and offered a video-first unified communications platform that spans video, voice, content sharing, and chat across desktop, mobile, and workspaces.

The platform grew more rapidly than its much larger competitors like Cisco Webex or Microsoft Skype because Zoom made things easy for its users. Easy to set up, use, and change one's background ... maximum simplicity, minimum effort. Zoom, for instance, is much higher quality in performance than Hangouts. It is easy to see why quality, the ability to record meetings, and the ability to share high-quality videos over that call help set Zoom apart from the competition.

Zoom offers far more robust application features than Skype does. Zoom allows one to hold meetings with up to 1,000 participants, while Skype limits to 100 people only. Zoom Chat is the glue holding unified communications together. It is the central place where users can connect instantly over instant messaging (IM), schedule or jumps into ad-hoc meetings and phone calls, share files, and keep track of tasks.

One does not need an account to sign on to a Zoom meeting. Since the platform is compatible with Mac, Windows, Linux, iOS, and Android, nearly anyone can access it. By April 2020, Zoom claimed it had 300 million daily meeting participants worldwide. In comparison, only six months before that, at the end of 2019, this number was only at 10 million meeting participants. At schools, Zoom is quickly becoming the go-to tool for teachers who teach online to their students because it has all these features that make for a dynamic and engaging classroom environment.

Chapter 21

From Central to Edge Cloud – Real-Time Intelligence and Solutions

This illustration shows a distributed hybrid cloud with both central and edge cloud vs. just a central architecture.

At the start of cloud computing services, a typical analogy is to make cloud computing similar to utility – namely, the cloud is centralized and distributed via a global network like the Internet similar to the power grid for electricity. Cloud is also utility based as a service, pay on demand and a subscription basis, etc.

Then as service applications advance, it is realized that the cloud has some key characteristics that are different from the general utility, and the two biggest of them are latency and security. Electricity, for example, is a power current, unlike a cloud that carries data or content, requiring a certain speed or low latency to function correctly and security as well to protect integrity. Indeed, the power grid needs protection, too, for its normal operations, but it is different from content security and data breach prevention.

For cloud traffic, latency is a mission-critical factor when we use the network to deliver content. Internet traffic can be typically categorized into two types: eyeball vs. content traffic. For example, when the general audience get online, our purpose is to look at things around and may consume a small

DOI: 10.1201/9781003305187-25

amount of data or video each day. But when a business-like Netflix kick off its video-on-demand online business, they distribute big chunks of data like a movie over the Internet daily per the viewer's demand.

For Netflix, the success of its business model depends on the quality of the video streaming and customer experiences. When watching a Netflix movie, if a viewer is experiencing delay, pause, and other interruptions, they may not use Netflix any longer. Therefore, latency must be reduced to the minimum for Netflix's success.

Let's say a Netflix content server is based in LA, California, while a viewer is sitting in Denver, Colorado; due to the relatively short distance between the two cities, when Netflix distributes a movie stream to Denver, latency may be less of a concern. But latency may become an issue if a viewer sits in New York City. What can Netflix do to resolve such issues? The easiest way is to create some so-called "proxy server" nearby NYC to shorten the distribution distance.

The proxy server will do magic called content caching. By deploying or using such a network of proxy servers, Netflix can distribute content from an "origin" server worldwide by caching content close to where each end user is accessing the Internet via a web-enabled device. By caching content physically close to where a user is and reducing the distance it has to travel, latency is reduced.

The above architecture is called CDN (also referring to Chapter 23) – content distributed network, which focuses on latency reduction for content delivery over the Internet or network. When cloud computing came into play, it made CDN deployment much easier.

In the past, a CDN would consume a lot of physical servers as proxy servers and accessories across the world to make things work, and the time to market was quite long. Now cloud computing can make such CDN setup virtually in the cloud and instantly on demand. CDN has become a standard feature of cloud computing networks, or more accurately, edge cloud across all local sites.

THE INNOVATIVE LEAP – BRINGING ICT TO THE EDGE

Enterprises today have more service requirements, driven mainly by IoT and other high-end remote IT applications such as telemedicine, smart agriculture, and warehouse robots. Such applications would need an instant response from cloud services nearby or the so-called edge computing, which also reduces the network traffic burden and central computing consumption. This is, of course, not saying centralized cloud computing is not needed; instead, it will make a hybrid model of edge and centralized cloud.

Edge computing is better suited for processing time-sensitive data, while standard cloud computing fits to process data that is less time driven but with more insights. Besides low latency, edge computing is preferred over standard cloud computing in remote locations because, on these sites, there is limited or no connectivity to a central cloud.

Edge computing creates new and improved models for industrial and enterprise-level business operations. This is critical and will maximize operational efficiency, improve performance and safety, automate all core business processes, and ensure "always on" availability.

The edge devices can include many different things, such as an IoT sensor, an employee's notebook computer, their latest smartphone, the security camera, or even the Internet-connected microwave oven in the office break room. Edge gateways themselves are considered edge devices within an edge-computing infrastructure.

NEW MINDSET FROM EDGE COMPUTING

Technical Outlook – Closer to the Applications

Edge computing is a new infrastructure paradigm that pushes computation closer to the data source. Edge cloud enables the execution of a broad range of devices, from bare metal to microservers. It enables faster connections, data preprocessing and cleansing, and reaction time compared to cloud services.

Key components of edge computing comprise compute CPU, local storage, data management, data analysis, and networking interface. Depending on the complexity of the workload, the computing infrastructure might range from a simple MCU (microcontroller unit) to a high-end GPU (graphic processing unit).

Edge computing is designed and deployed for intelligent devices locally. Sensors and other instruments are gathering and processing data – to expedite that processing before devices connect to IoT and send the data for further use by enterprise applications and personnel.

Business Outlook – Productivity from the Edge

To start with edge computing, think about the entire system and how the edge devices will fit into the larger architecture. Find the best devices for the design and the company's overall goals. Ask specific questions about the devices. How can they be maintained and upgraded?

Think about latency and workload as well. Edge cloud can achieve much faster response time (very low latency from when an event happens until a response occurs). Edge computing reduces the amount of data traveling over a wide area network. Keeping the workload local is also an obvious bonus from a security perspective.

Edge computing aims to bring business applications closer to where the data is created and action must happen. It is like local data centers for data storage and processing. Businesses now have reliable connectivity for their IoT applications, even when public cloud services are affected or have an outage. Edge computing would allow IoT applications to use less bandwidth and operate normally under conditions of limited connectivity.

Application Outlook – From Manufacturing to Healthcare

Edge computing is already used in many ways, from the wearable on your wrist to the computers parsing intersection traffic flow. Other examples include smart utility grid analysis, safety monitoring of oil rigs, streaming video optimization, and drone-enabled crop management.

Edge computing helps to move applications, data, and services away from centralized hubs and the cloud to the logical extremes of a network. It also empowers analytics and information generating from the data source.

For instance, multi-access edge computing (MEC) is a 4GLTE/5G network + edge cloud bundled solution that provides services and computing functions that apps require on edge nodes. It makes application services and content processing closer to users and implements network collaboration, providing users with a reliable and ultimate service experience.

MEC also allows network operators to open their networks to various innovative services. This gives rise to a brand-new ecosystem and a value chain. Furthermore, MEC, as an enabling technology, will provide new insights into the coherent integration of IoT in 5G wireless systems.

INSIGHTFUL PRACTICE

Digital Healthcare based on Edge Cloud

Most new opportunities for the "cloud" lie at the "edge." Edge cloud supports and collaborates with game-changing technologies such as IoT, artificial intelligence, and robotics, reshaping healthcare by pushing big data and storage closer to the source.

IoT gadgets can deliver tremendous measures of patient-generated Health Information (PGHD). Healthcare suppliers, thus, might access data about their patients continuously instead of dealing with moderate and fragmented databases.

Medical imaging apparently can take advantage of the cloud in healthcare since the cloud can store and share large data files involved in medical imaging, making an optimal solution. It saves hospitals, physicians, and other organizations in healthcare costs while boosting speed and efficiency.

Edge can support advanced remote-patient monitoring, by processing data received from medical devices such as glucose monitors and blood pressure cuffs. Then the reports will remind and alert clinicians to problematic readings. Advanced technologies are also finding ways to handle cloud security, such as identity management and access control, Internet-based access, authentication and authorization, and cyber criminals in healthcare cloud computing.

The synergy of IoT and edge cloud enables healthcare professionals to be more watchful and proactively connect with patients. IoT devices tagged with sensors are used for tracking the actual time location of medical equipment such as wheelchairs, defibrillators, nebulizers, oxygen pumps, and other monitoring equipment.

AI Virtual Assistant from Edge Cloud

Virtual assistants are typically cloud-based programs with devices and applications. As a virtual assistant serves the end user, the embedded AI uses advanced algorithms to learn from data input and work to better predict the end user's needs. AI assistants can perform simple jobs for users, such as adding tasks to a calendar; doing quick online searches and providing information; or controlling and checking the status of smart home devices, including cameras, lights, and thermostats.

An AI assistant can streamline and automate basic customer service or sales interactions. The birth of the first virtual assistant began with an IBM device named Simon in the early 1990s. It was a digital speech recognition function that later became a feature of the PC with IBM and Philips.

What is edge AI? AI relies heavily on data transmission and computation of complex machine learning algorithms. Edge computing sets up a new-age computing paradigm that moves AI and machine learning to where the data generation and analysis occur: the network's edge.

By joining edge computing architecture into their systems, organizations can improve performance and diminish inactivity. Instead of AI virtual assistants sending data and preparing requests to a concentrated server, they can disperse the weight among edge data centers nearby and play out some processing capacities locally.

Smart Manufacturing with Edge Cloud

Smart manufacturing can accurately predict requirements and identify errors. It makes innovation and the manufacturing process more manageable. It is the process of combining and implementing various technologies and solutions into the traditional manufacturing process. Smart manufacturing enables factory managers to collect and analyze data automatically. In this way, they make better-informed decisions and optimize production. Edge cloud and IoT ensure that the data obtained from sensors and machines are communicated to the cloud by IoT solutions deployed at the factory level.

These demands are responsive to cost, quality, value-added customization, environmental effectiveness, etc. Smart manufacturing enterprises adjust with more flexible production of variable volumes of products, become less vertically integrated, and are more information-driven. By joining a data warehouse and enrolling in industrial equipment, manufacturers can share data that will consider better designs and processes, allowing them to reduce costs and crucial usage while keeping up better constancy and beneficial uptime.

A smart factory has three key elements:

1. Robotics: An imperative of Smart Factory 4.0 is robotics – autonomous mobile robots (AMRs) can be deployed to realize smart material handling and adaptive operations
2. Big data and IoT
3. Cloud-based management

Edge computing enables manufacturers to filter out data to reduce the amount sent to a central server, on-site, or in a cloud. Such ability to monitor and manage their assets remotely helps manufacturers generate new revenue streams. Through IoT technology and edge computing, machines can seamlessly talk to each other and react to any problems. The enhancement in precision manufacturing by these smart technological systems lowers error rates and product failures, overall reducing huge costs historically faced by manufacturers.

This illustration shows with edge cloud, on-site robots can have more work done effectively and boost productivity.

Chapter 22

From Virtual Machines (VMs) to Cloud Containers – Agile Computing

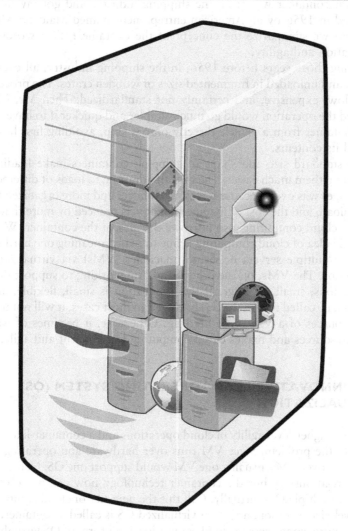

This is an example of one hardware to host multiple virtual machines which are actually software instances.

DOI: 10.1201/9781003305187-26

We often regard the cloud computing service model as the "utility model," as we use water and electricity at home. When we need it, we turn on the pipe or switch. When we are done, we shut it off – just this simple, convenient, and agile.

However, for cloud computing to become really agile, the mechanism behind the scene is more complex than water and electricity. A secret weapon for cloud agility is by using the container module, which is a relatively new term in the ICT world.

Most of us have seen containers in the transport and shipping industry, either by sea or land, sometimes even by air. Why do we use a container? The first container was from the shipping industry and got invented and patented in 1956 by an American entrepreneur named Malcolm McLean. How do we summarize the concept of the container? Two words: standardization and agility.

During those years before 1956, in the shipping industry, all cargo was loaded and unloaded in fragmented sizes of wooden crates. The process was very slow, expensive, and certainly not standardized. Then Mr. McLean realized the operation would go much simpler and quicker if to have and lift one container from a vehicle directly onto a ship, avoiding first having to unload its contents.

The standard size and shape of shipping containers make loading and offloading them much easier. The time of adjusting to loads of different sizes and shapes was over. Because it was much faster and more organized to load and unload, and the cost of loading freight was reduced by more than 90%.

Now cloud computing also borrows the idea of the container. We know the basic idea of cloud computing is one to many, meaning one hardware to support multiple servers or virtual machines (VMs) via virtualization or hypervisor. The VMs, on the other hand, are working to support different applications, small, or large. If the application is small, flexible, and on-demand, or called microservices, however, in most cases, it will not need the full resources of a VM to be fired up. Otherwise, it becomes of waste of cloud resources and makes cloud computing less efficient and agile.

THE INNOVATIVE LEAP – OPERATING SYSTEM (OS) VIRTUALIZATION

We need higher-level agility in cloud operations and a container-like solution to solve the problem. Since VM runs over hardware and operating system (OS) runs over VMs, usually, one VM would support one OS. Namely, it is a 1-to-1 relationship, but the container technology now can make one VM support multiple OSs virtually. Call this the next tier of cloud computing or OS-level virtualization, and each virtualized OS is called a container.

Therefore, containers in cloud computing are a form of OS virtualization. A single cloud container can be used to run a small microservice or software to a larger app. Inside a container are all the executable components required: binary code, libraries, and configuration files.

In a nutshell, the benefits of cloud containers are apparent:

- Less overhead: Unlike traditional or hardware VM environments, containers require fewer system resources because they do not include OS images.
- Increased portability: Smaller in size means portable in actions.
- More consistent operation: The apps become customized to the client's needs.
- Greater efficiency: Containers get the work done with less energy consumed.
- Better application development: Containers make microservices possible and make the developers' work easier to manage and thus more effective.

A side note is that although containerization offers many benefits, it will not wholly replace VMs. That is because containerization and VMs have particular capabilities that help solve different solutions.

NEW MINDSET FROM CLOUD CONTAINER

Technical Outlook – Portable Cloud Computing

The advantage of a cloud container is that it requires fewer IT resources to deploy, run, and manage. A single system can host many more containers as compared to VMs. The fine-tuned mini architecture makes it possible to package software and its dependencies in an isolated unit, namely, a container, which can run consistently in any environment.

Containers do not use a hypervisor. Instead, they share the host OS's kernel. Hence, they avoid the infrastructure overhead of a full-blown OS. In this way, they provide only those resources (i.e., installations, dependencies, and code) that your applications need.

Containers as software can virtually package and isolate applications for deployment, abstracting at the app layer with code and dependencies. Containerized applications can be composed of multiple container images and containers run faster than VMs because they have less overhead.

Business Outlook – Product Development across Platforms

Containers bring about a seamless development process due to their capacity to provide portability between multiple platforms and clouds. It has outstanding efficiency and can deliver higher utilization of resources. With containers, the development-and-deployment process shrink to minutes with its microservice-based architecture.

The combination of containers with microservice accelerates product development and launch and enables a faster go-to-market. Software DevOps is always about better product delivery. Container orchestration has contributed significantly to this goal. Overall, containerization is an innovation that helps stay ahead of the software development curve. It keeps pace with the ever-changing demands and standards of the market.

As the market has no looking back regarding technical advances, containerization opens the door to various opportunities to help your customers with multiple benefits like efficiency, portability, speed, scalability, improved security, and agility.

Application Outlook – Microservices and CaaS

So, containers are thus exceptionally "light," meaning they are lighter versions and more portable than VMs. In deployment, they are only megabytes in size and take just seconds to kick off, vs. gigabytes and minutes for a VM. Containers also reduce management overhead.

Containers as a service (CaaS) allows software developers and IT departments to upload, run, scale, organize, and manage containers. Please note that CaaS is different from the PaaS (platform as a service) since it relies on containers. CaaS also provides edge containers and decentralized computing resources as close as possible to the end user to perform connectivity reliably, better manage the data, reduce latency, save bandwidth, strengthen security, and enhance the overall digital experience.

CaaS includes microservices which is an architectural design for building a distributed application. Microservices are an architectural paradigm, while containers are a means to implement that paradigm. Containers host the individual microservices that form a microservices application. Microservices break an application into independent, loosely coupled, and individually deployable services.

INSIGHTFUL PRACTICE

Microservices for Netflix

In Chapter 9, we talked about Netflix using AWS. In particular, Netflix uses a microservice architecture on AWS. Microservice architecture is a method of development where a large application is divided into small modular services. It enables an organization to scale its workload with ease. It also maintains a cost-effective operation in the cloud and avoids a single source of failure even if engineers change multiple service areas across the board.

A microservice has a single function, such as making an online payment, routing network traffic, or analyzing a medical result. The adoption of microservices applications is increasing exponentially. Microservices run on easily packaged, lightweight containers and are designed to run anywhere while multiple containers can be deployed in a single VM.

Netflix is one of the first major drivers behind microservice architecture. By adopting microservices, Netflix engineers easily change any services, leading to faster deployments. More importantly, they can track the performance of each service and quickly isolate its issues from other running services.

Microservices use modular structure, which is particularly important for larger teams handling large products. Such independent deployment brings about simple services that are easier to deploy. They are autonomous and are less likely to cause system failures when they go wrong.

From a performance perspective, containers make a much better foundation for microservice architectures than VMs. With their small size, containers are also much more efficient at start off than VMs as they do not require the OS start time. Containers, therefore, are the perfect place to run microservice-based applications.

IoT over Container Kubernetes

Many IoT systems carry large numbers of sensors to collect data, which need to be processed for intelligent decision-making. Cloud is essential for aggregating data and gaining insights from that data, as well as high scalability. One benefit of containers for IoT is that developers can create container development-and-deployment environments that enforce specific work and testing processes before they deploy applications into production environments.

Containers, as a cloud technology, allow the management of IoT devices and sensors with agility; the devices can be easily scaled up and down. Hence IoT App development gets an agility, productivity and security boost from container technologies.

Kubernetes, as an example, is an open-source container management platform, and it unifies a cluster of machines into a single pool of computing resources. With Kubernetes, you organize your applications in groups of containers, which run using the Docker engine on every single node, keeping your application running as you request.

Kubernetes is the container trigger and helps developers quickly verify, launch, and deploy changes to IoT services. Also, Kubernetes is supported by most cloud providers, such as AWS, Azure, and Google Cloud. It is even supported as on-prem software and is easy to migrate to any cloud platform in the future.

Containers for Multi-Cloud

Multicloud is a cloud approach comprised of multiple cloud services from more than one cloud vendor – public or private. Companies use multi-cloud environments as diversification to distribute computing resources. Multiclouds also minimize the risk of downtime and data loss. They can also increase the computing power and storage available to a business.

The container runs multiple processes that are isolated individually from the rest of the system. Hence containers are highly portable and stay consistent from development to testing to production. They also help reduce issues between development and operations teams by separating responsibility. Now operations can focus on infrastructure, and developers can focus on apps.

As mentioned about microservices, containers deliver both abilities as well as portability. Developers can move data between the edge and the public cloud via data stores or services within containers. This portability is also beneficial for app testing. Developers can test their applications across multiple OSes. This can easily demonstrate if that makes a difference to any results. In case an application crash during testing, it only affects one specific container instead of the entire OS.

It is called container clustering when combining multiple containers as a team of microservices. Admins can more easily scale container systems, meanwhile increasing their resilience. Individual services can be updated without impacting the whole application. Using containers in the multicloud envitonment can also benefit future innovations for existing applications: speeding up innovation, DevOps implementations, and agility to application development.

This illustration shows cloud computing can achieve better agility if it can scale via containers with the computing needs from each service or application.

Chapter 23

From Web Surfing to Over-the-Top Streaming – Video and Gaming Frenzy

This is an example of using a smartphone to do live video streaming.

To many people, the Internet is the tool and media via which we can connect and run communications, such as sending emails and chats, hanging around on social media, browsing web portals, and even doing some eCommerce. All this is called eyeball traffic, while large file transfer, video, and streaming are called content traffic. As we know, traditionally, video broadcasting was accessible on a television set via satellite or cable. But the 4th Industrial Revolution requires the Internet can also be used to transport any content, including real-time streams, video clips, and voice communications.

The Internet originally was not developed to send content traffic like videos and streaming. To make Internet transferring video smoothly, we need an extra Internet layer called CDN (content distributed network, also refer to Chapter 21) – a network and deployment of servers that distribute

DOI: 10.1201/9781003305187-27

content from an "origin" server throughout the world by caching content close to where each end user is accessing the Internet via a web-enabled device. CDN leverages a worldwide network of servers at the edge and disseminates content from various interconnected hosts. The content they request is stored on the origin server and replicated and stored elsewhere as needed.

CND makes OTT (Over-the-Top) possible, which is the software and equipment that bypass traditional streaming technology like TV, and broadcast video content over the Internet at the request to suit the individual consumer's requirements. OTT initially referred to devices that go "over" or bypass a cable box and directly give the user access to TV content.

OTT content refers to, in most cases, film or TV content on a phone, laptop, tablet, or connected TV. Consumers can stream OTT content anywhere, at any time – a big perk over traditional TV programming. As we may watch daily, some of the most popular OTT providers include Netflix, Amazon Prime Video, and Hulu.

THE INNOVATIVE LEAP – CONTENT OVER EDGE + CDN

Content is delivered and distributed to geographic locations closer to users through local points of presence (PoPs) at the edge. When requested content is cached (pre-saved) by a CDN's servers, an end user's Internet Service Provider (ISP) or mobile provider gets that content by connecting to a server on the CDN's network rather than waiting for their request to go a long way to the origin.

Like any technology, CDNs also must evolve to keep up with heavier Internet traffic nowadays. This includes a broader variety of content – streaming video and gaming. For younger users, such traffic demands ultra-low latency, increased quality of experience, and different types of connected devices. All these demands require strategic innovation, which can be achieved by using edge computing to CDNs.

Be aware that the major difference is that a legacy CDN focuses mostly on data delivery; on the other hand, edge computing platforms can offer various additional digital services with intelligence.

An edge platform serves far beyond just content distribution because today's content-heavy applications require advanced intelligence, analytics, and optimization capabilities. While edge cloud provides the speed, resilience, demand, and connectivity of modern times, it is also a new technology going hand in hand with the latest technologies, such as 5G, AR/VR, AI, and IoT.

NEW MINDSET FROM OVER-THE-TOP

Technical Outlook – Edge Cloud-powered CDN

An edge location is initially designed for cloud computing users to access the cloud services they subscribe to from vendors like AWS or Azure. They are located in most major cities worldwide and are now also used explicitly by CDN vendors like CloudFront to distribute content to end users to reduce latency and boost the services from OTT firms.

Edge-powered CDN platform can make a big difference for content distributions online. It provides ultra-low latency and globally distributed edge locations; with a big reduction in infrastructure and connectivity costs, it increases network capacity and avoids bottlenecks. All this allows greater scalability and flexibility for transferring content globally, even during peak hours.

An edge platform can also bring scalability, intelligence, and greater security to the service providers and end users. Intelligence can be programmed and applied directly at the edge and in real-time. By combining edge computing with CDN nodes, applications can run at their highest level of performance with global coverage. Overall, its added computing power enables more control and new use cases.

Business Outlook – Transforming Customer Experiences

OTT platforms allow users to access media services anytime and anywhere. And this stands as the biggest advantage of OTT platforms. Nowadays, nearly half of the people watch TV on devices other than an actual TV. The key drivers of adopting and using OTT video streaming platforms are performance expectancy, price value, habit, and content availability.

Again, the purpose of CDN is to help improve the load time of a website by using the closest CDN server. For example, without CDN, someone in Asia visiting a web server in the United States must wait for the traffic to travel the round trip. A CDN can shorten the path information has to travel between the server and the users because it is a network of servers deployed in multiple geographic locations. When a user accesses a website, information is pulled from the server closest to the user with a faster data flow.

With digitalization, people will often use OTT platforms as their mode of entertainment. For content creators and developers like movie and TV makers, online platforms can provide the capability to grow and gain popularity and a great place to expand and gain expertise.

Application Outlook – Online Streaming Explosion

In July 2022, streaming platforms outpaced broadcast and cable in total TV viewership in the United States (source: https://www.npr.org/2022/08/18/

1118203023/streaming-cable-broadcast-tv). This is a historical record. To stream OTT, customers require a high-speed online connection and a device that supports apps or browsers. Via mobile OTT devices, such as Smartphones and tablets, viewers can download OTT apps to stream on the go. Via PCs, consumers can access OTT content from desktop-based apps or web browsers.

To meet the requirement for a digital application to deliver modern content , OTT programs need global scalability, distributed security, high resiliency, and minimal latency. Otherwise, OTT-related applications for business may be doomed to fall behind. Plus, APIs can simplify integration between systems and configurations and bring benefits closer to your users with leading-edge computing technology.

For individual users, live streaming platforms provide video hosting that allows users to upload and broadcast videos to their audience. Nowadays enterprises can rely on edge locations with global coverage and 24/7 support, and use online video platforms to share videos for lead generation, brand awareness, advertising, and providing paid access to video streaming.

INSIGHTFUL PRACTICE

Online Streaming and OTT

Again OTT sends content over a high-speed Internet connection. This replaces traditional distributors such as broadcasters, cable, and IPTV operators sending the content. OTT content can be watched on a phone, laptop, tablet, or connected TV.

Digitization makes video over the Internet possible; just like routine data sent over the Internet, audio and video data can be broken down into data packets and sent online. Each packet carries a small piece of the file, and together comes the flow of packets. Then an audio or video player on the client device takes the flow of data packets and converts them as video or audio.

In some real-time events, speed is far more critical than reliability for online streaming. For instance, video conferences prefer interacting with the other conference attendees in real-time rather than experiencing delays. Therefore, a few lost data packets may not be a huge concern, and speed comes first.

Streaming allows users to enjoy video watching continuously online. Streaming media players load a few seconds of the stream ahead of time, so the video or audio can continue playing if the connection is briefly interrupted. This is known as buffering, ensuring that videos can play smoothly and continuously. However, over slow connections, or if a network has a great deal of latency, a video can take a long time to buffer.

Two major benefits of OTT Platforms are: First, cost-effective – anyone willing to watch online content can watch for free or register for the service with a monthly or yearly subscription. Second, easy access – people can log in to these OTT platforms through mobile phones, tablets, laptops, etc.

The YouTube Story

YouTube is a free video hosting platform, and its algorithms control the platform's daily operation. Originally launched in 2005, YouTube quickly became the online video content champion. It is attracting over one billion regular users globally. The Google-owned portal gained popularity by enabling people to share their videos with others worldwide: the polatform encourages viewers to express their opinion via videos they create or watch, store videos they like, and share them with friends and family members.

This highlights YouTube – the very popular online video portal. YouTube is an over-the-top application.

YouTube videos cover a variety of topics anyone cares to upload a video about. The video can be an amusing clip of their pet or footage of them traveling worldwide. The popularity of YouTube's success is mainly due to its ease of use. Videos in different file formats can be uploaded to the portal. YouTube then converts the uploaded files into its Adobe Flash video format. This enables the video to be played on YouTube and can be installed on users' computers or smart devices for free.

YouTube also allows the embedding of videos on other websites or content formats by simply copying and pasting the URL links. People can watch a video on a website using a YouTube player. This saves the trouble of hosting the video on a website that requires a lot of bandwidth. As introduced previously, bandwidth is the pipe size or range of signal frequencies needed to transmit data over the Internet at the cost of the amount you use. YouTube, now a video host portal, carries the bandwidth weight for other sites that want to display video.

Today, YouTube has attracted 2.3 billion users worldwide. People watch one billion hours of YouTube daily, 8.4 minutes per day per user (source: https://www.oberlo.com/blog/youtube-statistics).

OTT Boosting Online Advertising

We expect US subscription OTT video ad spending to near $10 billion and account for 3.4% of all digital ad spending—and 10.2% of total video ad spending—by the end of 2023 (source: https://www.insiderintelligence.com/content/us-subscription-over-the-top-connected-tv-advertising-won-t-slow-down-next-year).

The main reason OTT platforms have become so popular is due to the online open platform; viewers, especially the younger generation, can directly access these platforms on the go. Since one can access these platforms from their mobile or laptop, the person does not need to find a television to watch their favorite TV shows. All the person needs is a stable internet connection, and it is good to go.

The rise of OTT has driven business trends to provide better networks and content, which can maximize digital connectivity and traffic to its core. As the OTT platforms are getting more popular among the audience, advertisers keep rolling out their products' showcase to obtain quality traffic and a sizable audience. This helps to increase their marketing Return on Investment (ROIs) and higher conversion rates based on different user behaviors and preferences.

With higher data analytics technologies and capabilities, OTT platforms also provide advertisers the gain of analyzing who their audiences are, what they need, and how their tastes may be catered to. This definitely helps facilitate higher conversion rates. Thus, micro-targeting gets easier, and there is additionally more engagement with the customer.

Chapter 24

From Data Center to Hyperscaler – Pivotal Data Power

This figure shows the holistic view of modern hyperscale data centers.

Digitalization is all about data and how we better use data to change lives, work, and the world. A big question is: where are the data stored and processed? The answer can be two-folded: data are generated from all over the place, and data are processed and held mostly in data centers (DCs) or virtual DCs, namely, the cloud.

A DC is a physical site that organizations, including businesses and agencies, use to house their critical ICT applications and data. A DC is designed for a computing and storage network that enables the delivery of shared applications and data. When data keep growing, organizations need a centralized space (therefore the name DC) to house and maintain back-end ICT systems and data stores – its mainframes, servers, router, switches, and databases.

What is a DC made of? DCs have three primary components: compute, storage, and network. DC devices include routers, switches, firewalls, storage

DOI: 10.1201/9781003305187-28

systems, servers, network connectivity infrastructure, and application delivery controllers. Since DC components store and manage business-critical data and applications, DC security is also critical in DC design.

Some time ago, communications mainly were about voice services such as local, long-distance, international, and mobile. Not too many data communications were going on. Hence, we did not hear much about DCs. Then Internet started to change the whole landscape, shifting the ICT operations toward being data-centric.

When cloud computing came to the trend back in 2009, DCs became the powerhouse of digitalization, namely, as a power plant produces and distributes electricity, a water plant produces and distributes drinking water, and a DC now produces and distributes cloud computing.

DCs can be onsite or remote. An onsite DC is typical when an organization only has a few offices. But when an organization has multiple working sites, keeping the DC in the headquarter (HQ) site or a remote site with common access from all branch offices makes better sense.

THE INNOVATIVE LEAP – EMERGING SUPER DATA CENTER

DC tiers are an efficient and standard way to describe the infrastructure components of a business's DC. A Tier 4 DC is designed as more complex and robust than a Tier 1 DC, but this does not necessarily mean a higher tier is best suited for a business's needs. Using Tier 1 infrastructure might leave a company open to risk; Tier 4 infrastructure, however, might be an over-investment. So, it all depends on the business needs.

The Four Tiers of Data Center

Tier 1 data center	Has a single setup for power and cooling. This means it does not have redundant and backup components. As a result, it has an expected uptime of 99.671%, or 28.8 h of downtime, to manage annually.
Tier 2 data center	Has a single setup for power and cooling but with some redundant backup components. It has an expected uptime of 99.741%, or 22 h of downtime, to handle annually.
Tier 3 data center	Has power and cooling systems redundancy. So the data center updating and maintenance can happen without taking it offline. Such a data center has an expected uptime of 99.982%, or only 1.6 ho of downtime to worry about annually.
Tier 4 data center	Being fault-tolerant, has complete redundancy for every component. It has an expected uptime of 99.995% or just 26.3 min of downtime annually.

Beyond Tier 4 DC, we now also have so-called hyperscale DCs, which are massive business-critical facilities designed. Such DCs will efficiently support robust and scalable applications that are often associated with big data-producing and processing companies such as Google, Amazon, Facebook, IBM, and Microsoft. These companies require huge amounts of space and power for their DCs. This way, they can support massive scaling across their cloud, big data, storage, and analytics platforms.

As of 2022, there are over 700 hyperscale DCs globally. Most 39% of global hyperscale DCs are built and located in the United States. China takes second and Japan third place in hyperscale DC operations, respectively.

As for specifications, so far, the DC industry has no minimum specification for hyperscale DCs. In general, such DCs are built to house robust, scalable applications, and storage-draining solutions for cloud and big data storage. The facility features can include around 500 cabinets upwards, a minimum of 5,000 servers, and at least 10,000 square feet in floor space with an ultra-high speed, high fiber count network.

NEW MINDSET FROM HYPERSCALE DATA CENTER

Technical Outlook – Scaling with Data Growth

What is a hyperscaler? This term stems from hyperscale computing, an agile method of processing data. Depending on data traffic, the scale can quickly go up or down. Hyperscalers have applied this computing method to DCs and the cloud to handle fluctuating demand.

As its name implies, hyperscale is for achieving massive scale in computing – typical characteristics of big data or cloud computing. Hyperscale infrastructure is designed for high scalability, leading to outstanding performance levels, throughput, and redundancy. Such DC features enable fault tolerance and high availability.

DCs are energy-intensive buildings because their size and number must meet the increasing demands of a digital economy. Hyper-convergence combines computing, storage, and networking into a single system. Such innovation helps reduce DC complexity and increase scalability.

DCs going hyperscaler would call for massive processing power, automation, and intelligence. The main differentiators in today's market for DCs will be scale, AI-powered applications, smart monitoring tools, and efficiencies. These will drive cost and power savings.

Business Outlook – Becoming Cloud Native

Hyperscale DC is all about cloud use and apps. Now businesses realize the dynamism of what can be done over the cloud with their data. They are migrating from their existing resources on-premise to well-equipped DCs and the cloud for better data management. DCs and the cloud have become a top priority for businesses across the globe to measure up their ICT infrastructure requirements.

DCs drive growth, generate employment, and boost the digital economy. According to a research firm RTI International (source: https://www.rti.org/), for one DC function or role, five more jobs are associated elsewhere in the economy from the field of power, device, security, and networking.

The adoption of enterprise cloud-based technologies and consumer video applications is driving DC explosion. With DCs rapidly proliferating around the globe, they reduce latency and improve application performance. Plus, edge DCs also have a substantial effect on network demands and topology.

Application Outlook – For MetaAlverse and Automation

Hyperscaler DC can be applied to cloud computing, big data, AI apps, distributed systems, and companies. Typically, off-site DCs handle a hyperscale cloud, which is known as the public cloud.

We know cloud companies have DCs and run their cloud from these DCs. Organizations often have their own or on-premise DCs as well. With the cloud, data are stored with scalability, and applications run off-premises and are accessed remotely through the Internet.

Do we still need physical DCs in the era of cloud computing? Of course, the cloud is operated from those Tier 3, Tier 4, or hyperscaler DCs. But the legacy on-premise DCs is quickly getting obsolete since the trend is to run IT operations from the cloud from off-premise DCs. In a nutshell, DC will forever exist but be a matter where they are. DC and cloud make the powerhouse of digitalization.

INSIGHTFUL PRACTICE

The Equinix Story

Equinix is a global digital infrastructure company comprising 200+ DCs and cloud computing. It interconnects industry-leading organizations in finance, manufacturing, mobility, transportation, government, healthcare, and education across a cloud-first world.

Founded in 1998 by two facilities managers at the former Digital Equipment Corp., Equinix is the pioneer of some of the first DCs in the USA. The centers served as connection points for networks forming the early Internet.

Today Equinix DCs are in 31 key markets across the Americas, EMEA, and Asia-Pacific regions; they offer more than 7 million square feet of space, making a global network of high-performance DCs maintaining a 99.999% uptime record. This standard is a key factor for the world's most demanding organizations. Today Equinix owns and operates over 220 International Business Exchange™ (IBX®) DCs in 63 major metros across the globe to make interconnection easy.

Equinix's power comes from its ecosystem. Its clients are Amazon, Apple, AT&T, Facebook, and Nokia. The reason major enterprise customers go with Equinix is that within Equinix's DCs, there are lots of carriers and network providers (source: https://www.equinix.com/).

The core part of Equinix's business model that generates most of its revenue is through "colocation," or providing space, power, cooling, and security for its customers' servers at Equinix's own DC locations. Today, Equinix connects more than 4,000 companies to their customers and partners inside their networked DCs. Equinix enables customers to keep pace with the fast-changing digital marketplace.

SD-Data Center – DC as a Service

An SDDC (software-defined DC) makes all its infrastructure elements – networking, storage, CPU, and security – virtualized and delivered as a service. Deployment, operation, provisioning, and configuration are abstracted from the hardware. The two major technology drivers behind SD-DC are virtualization and software-defined networks (SDN).

Do not mix SDDC with cloud computing. SDDC enables infrastructure automation like for how to facilitate cloud with a focus on individual infrastructure elements, while a cloud operating model allows IT services to be consumed by clients based on application requirements. SDDC is cloud service provider (CSP) facing, while cloud is end-user facing.

The primary benefit SDDC brings about is agility. With an SDDC infrastructure, CSPs can allow users to reduce the time to provision new ICT resources significantly. It now only takes hours or minutes to setup a new physical server, provide scalable storage capacity to an application or modify physical networking.

A SDDC differs from a private cloud since a private cloud only has to offer virtual-machine self-service, beneath which it could use traditional provisioning and management. Instead, SDDC concepts imagine a DC encompassing private, public, and hybrid clouds.

DCs have traditionally been 'hardware-centric" and focused and relied on physical equipment. With new technologies like virtualization, all primary services in a DC can be virtualized. Pioneered by VMware, the SDDC extends virtualization beyond computing (i.e., servers) to network and storage.

Unsurprisingly, the core components of an SDDC consist of network virtualization, server virtualization, and storage virtualization. It also has a business logic layer to translate application requirements, SLAs, policies, and cost considerations.

Google Hyperscale DC Story

In April 2022, as demand for Google Cloud soars, Google announced it would build a new DC in Nebraska consisting of four buildings totaling more than 1.4 million square feet. This is part of Google's ongoing strategy to invest billions yearly in constructing and equipping hyperscale DCs to meet its growing cloud customer demands.

Google DCs are vital in storing the world's information and making it universally accessible. Google owns and operates the world's largest fleet of DCs that keeps the Internet running and Google products accessible to consumers and businesses globally 24/7/365. Google already had 21 major DC locations by 2020 in the United States. In 2022 Google plans to invest $9.5 billion in US offices and data centers around the country, up from more than $7 billion in 2021 (source: https://www.google.com/about/datacenters/).

The DCs allow Google to provide a search engine, Gmail, YouTube, cloud-computing products like Google Cloud Platform (GCP), and other web-based services such as Google Workspace that are vital for economic activity. Google DCs fall under the category of hyperscale DCs.

Google's relentless focus on innovation resulted in its DCs being among the world's most high-performing, secure, reliable, and efficient. Google has set the aim high with its innovative uses of technologies. The company (founded in 1998 with green initiatives) announced last year that its global information empire would run entirely on carbon-free energy by 2030. Google became the first carbon-neutral Internet company in 2007.

Google and other hyperscale computing vendors are reshaping how data flows worldwide; they reroute undersea cables to target emerging markets, bringing traffic ashore near their big enterprise users and consumers globally in the booming cloud-computing market.

This shows some insight of data center setup with many server cabinets.

Chapter 25

From Service Chaining to SASE – Solution Fabric Stitching

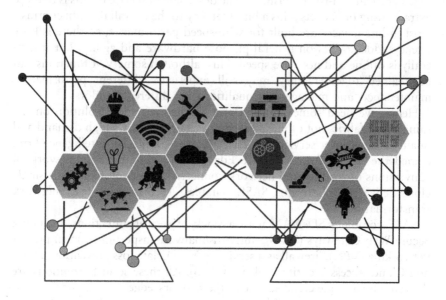

This illustration carries the idea to offer better communication services over a single stitched fabric or solution stack.

Digitalization goes beyond many so-called traditional "silo thinking or controlling" models, providing a central platform to converge different technologies. For instance, traditionally, network service is different from security service. And application management differs from network management. Network focuses on traffic transport, security focuses on guarding on prem, and the application ends are user-oriented.

Now digital platforms (refer to Chapter 20) and cloud computing provide great opportunities to converge different technologies and offer holistic solutions. For instance, SD-WAN (refer to Chapter 8) is a next-generation network solution powered by SDN and NFV technologies (refer to Chapter 4). SD-WAN can directly connect to the cloud, Internet, and mobile end

DOI: 10.1201/9781003305187-29

users, route traffic based on application priorities and policies, and automatically select the best routes.

But a native SD-WAN solution only comes with basic traffic encryption as security protection. At the same time, as it provides more connection options and service domains, it also exposes the network, network service providers, and end users to more security risks. At the start of SD-WAN, when the client needed better and more robust security services, the vendors often used a method called "service chaining" via APIs (refer to Chapter 4).

Service chaining is an offline concept wherein multiple network services instantiated in software and running on x86 hardware are linked or chained together in an end-to-end fashion. The downside of service chains is that the extra routing of packets adds a bit of latency to the overall flow. In contrast, most in-line devices are built for wire-speed performance, so there is little latency added. VMs on general-purpose hardware add agility, but it generally is offline and not wire speed. Plus, although some VM functions can be chained together, they may still experience different performance, maintenance, and management qualities from different vendors.

Hence, we need some "built-in and stitched together" solution on the same single fabric, not offline services, to cater to today's high demand and qualify for end-user services. In 2019, Gartner, a market intelligence firm, coined the Secure Access Service Edge (SASE) concept, a framework of conversions of security and network connectivity technologies into a single cloud-delivered platform. SASE enables secure and fast network transformation.

SASE combines SD-WAN with a stack of security functions, including Secure Web Gateways (SWG), antivirus/malware inspection, virtual private networking (VPN), firewall as a service (FWaaS), data loss prevention (DLP), and Cloud Access Security Brokers (CASB). All these security solutions are delivered by a single cloud service at the network edge.

THE INNOVATION LEAP – A SINGLE FABRIC OF SOLUTIONS

The SASE security model can help clients in several ways of flexibility. With a cloud-based infrastructure, SASE can implement and deliver security services such as threat prevention, sandboxing, DNS security, credential theft prevention, DLP, web filtering, and next-generation firewall policies.

A SASE architecture embeds all network and security capabilities in a single software stack reducing capital investment. It also frees IT staff to focus on strategic work, enables a coherent security policy deployment, reduces hardware complexity and cost, and moves the enterprise toward the on-demand, pay-as-you-go model.

SASE will transform multiple solution categories such as SD-WAN, Next-Gen Firewalls, SWG, WAN Optimization, CASB, and Zero Trust Network

Access (ZTNA). Enterprise leaders should understand this critical trend and take advantage of the benefits from the next-gen network and security architecture.

NEW MINDSET FROM SASE

Technical Outlook – SASE Orchestration

A well-architected SASE solution delivers more than just a stack of networking and security components. Please note that "chained together" point solutions are not SASE, whether hosted in the cloud or not. It would take a tight integration for SASE to improve user experience and reduce management complexity across services.

For instance, all the necessary classification and policy enforcement actions across multiple services – from next-generation firewall to SD-WAN to SWG are performed without having to decrypt and re-encrypt at each service, thereby minimizing latency.

Tighter integration enables to achieve of consistent visibility and policy enforcement across services. Only when we have common visibility using consistent terms and contexts can we assess what is happening in the environment, figure out what to do, implement consistent policies, troubleshoot problems faster and more accurately, etc.

Business Outlook – Plan Big and Start Small

SASE is a next-gen network and security solution combination. There are two main differences between SASE and SD-WAN: security and management. While some SD-WAN solutions may integrate a full security stack, they are not SASE at all.

SASE creates easier management and reduces the costs of multiple separate services. An organization has an integrated security stack vs. different networking security solutions. A single suite of security capabilities managed by a unified solution can also deliver better threat detection and data protection.

Plan SASE strategically because it should be a holistic solution. Evaluate how the SASE solution supports your ZTNA plans and implementation. SASE should be an integral part of your cloud migration plans. Start SASE small from the easy parts. Select a SASE solution with multitenancy capability and role-based policy settings; also highlight interoperability which allows integration between the SASE with 3rd-party solutions and on prem capabilities, offering consistency between cloud and on-prem configurations.

Finally, pilot and test the SASE service with specific goals for security capabilities, conversion, integration, and performance, before officially deploying the service across the board.

Application Outlook – Performance and Security Win–Win

SASE improves performance and minimizes latency to optimize the user experience. The single-pass parallel processing approach to applying security controls improves application and network performance, eliminating backhauled traffic flows that reduces client-to-cloud latency.

Top SASE use cases include rapid response to network demand. The Covid-19 pandemic of 2020 revealed why businesses need quick response to crises and emergencies, support for mobile users because SASE follows users, not offices, and VPN replacement due to SASE has many advantages over legacy VPNs.

As mentioned, unlike the traditional server-based VPN, SASE is offered as a cloud service. Thus, as with other SaaS solutions, you do not need to worry about the operation or maintenance of the underlying infrastructure. SASE provides controlled move-to-cloud services with easy traffic segmentation and handles cloud migration individually.

SASE is OS agnostic and can support multiple operating systems from the edge. SASE also guarantees identity-driven security: for every network connection, user and resource identities determine the level of access allowed, networking priority, and quality of service based on a unified organizational policy.

INSIGHTFUL PRACTICE

The Zscaler Story – Cloud-based Security Solutions for SASE

Zscaler is a cloud-based information security platform, functioning as a cloud-based proxy and firewall, routing all traffic through its security software to apply corporate and security policies. Its service is delivered through over 100 global data centers.

Zscaler offers a revolutionary security paradigm called Security as a service (SEaaS) that moves the security stack to the cloud. SEaaS protects all users with policy-based access and inline protection from malware and other threats, enabling local breakouts with full security controls. Zscaler's SEaaS includes such features as automatic visual (AV) inspection, Next-Gen Firewall, Sandbox, Advanced Threat Protection, URL Filters, SSL (Secure Sockets Layer) Inspection, and more – all in a unified platform service.

Zscaler's trademark – its Cloud DLP, using an advanced deep packet inspection engine, works as the core platform between users and the Internet. It inspects all traffic and provides the same comprehensive protection to all your users, no matter where they go – protecting your most vulnerable protocols.

Zscaler has partnered with major cloud service providers such as AWS, Google cloud – GCP, and Microsoft Azure. In this way, it delivers a network-agnostic zero-trust fabric of cybersecurity. Due to its offer of cloud native security services, Zescaler also becomes a partner for many SD-WAN vendors to forge the SASE solution to the enterprise clients nowadays.

This illustration is about the cloud base security concept. The advantages of cloud-based security include central control, real-time patching and upgrade, and easy backup and disaster recovery.

SASE as VPN Replacement

As mentioned briefly above, if you work from home or remotely, you most likely have the experience of using a VPN. A VPN tunnel sets an encrypted link between the user's device and a VPN server. It is uncrackable without a cryptographic key, so neither hackers nor your Internet Service Provider (ISP) could gain access to the data. This protects users from attacks and hides what they are doing online.

But VPN, as a legacy server-based on-prem technology, has many disadvantages nowadays. Now SASE is creating a global private network from the cloud for your company, replacing the legacy VPN. SASE and VPN operate differently. Whereas VPN is a standalone tool, SASE combines a number of platforms into one. For example, SASE incorporates services delivered through a cloud-based model such as a SD-WAN.

Again, SASE synchronizes networking and network security services over a single cloud-based platform. Traditional VPN was used over the public Internet for remote access to the internal data center, but it is not optimized for cloud access.

Plus, traditional VPN models rely on backhauling traffic to a centralized hub for security inspection. Today, with many business-critical applications powered by SaaS, data center-centric network infrastructure imposes a considerable performance penalty on applications and degrades the overall user experience.

SASE can deliver performance optimization compared to VPN as its platform includes a private backbone and built-in WAN optimization, avoiding the unpredictable Internet when connecting remote users to applications. This ensures that application performance from the remote is the same as from the office.

Zero Trust Network Access

Zero Trust is a concept that has been around for over a decade at least. It means: you only trust after verifying things constantly. Do not trust what you do not need to trust. Zero Trust projects can focus on networks, users, devices, or servers.

Based on the zero trust concept, ZTNA is developed as a security service that creates a context-based identity and logical access security around an application or set of applications. The access to a hidden app is restricted via a trusted broker to a set of named entities. Hence, ZTNA can provide secure remote access to an organization's applications, data, and services based on clearly defined access control policies.

Zero Trust improves mission outcomes by providing a long-term security posture to the enterprise, enabling capability advancements that were once unthinkable. When Zero Trust security is in place, you can provide protection anywhere on whatever device your colleagues choose. You can strengthen security further by including access management in the Zero Trust architecture to create a Zero Trust extended ecosystem.

SASE with Zero Trust thus supports a new and revolutionary approach: a preventive security strategy at the edge. The model runs a security inspection at the service edge, by understanding the identity of the users based on their access identity and enabling you to assign trust based on where they are in that network topology.

Chapter 26

From Street Malls to eCommerce – Digital Market Makers

This shows the eCommerce concept of completing transactions online.

eCommerce refers to the online commercial activities conducted and transactions over the Internet. B2C eCommerce means a business sells and provides merchandise or service to an individual consumer. Some examples of B2C eCommerce operations include Amazon.com, Walmart's online stores, and Alibaba's TMall.

The biggest advantage eCommerce enjoys is that it is online or over the Internet, which can reach far more customers than traditional brick-and-mortar stores and shopping malls; it shortens the lead time of purchasing and transactions for both sellers' and buyers' convenience. So, this is the exciting rather than challenging part of eCommerce for many people to envision.

DOI: 10.1201/9781003305187-30

What is challenging is to envision, establish, and manage the complete eCommerce ecosystem from supplies, and sales, to the transaction, payment, delivery, and post-sales customer management. You cannot fail in any of these procedures here. Otherwise, eCommerce will not work. For instance, many eCommerce firms rush up online and are eager to start selling. But if your payment procedure is hard to follow, your delivery is slow and poor quality, and your post-sales customer service is rude and bad, then the business will not fly.

One important thing worth noting is that eCommerce itself does not produce or manufacture anything. Hence, some call it a "virtual economy," and it makes only a key section of the whole industry value chain. On the other hand, eCommerce can help stimulate and increase production and manufacturing, which is called the "physical economy." For instance, when people place more smartphone orders online, the factories will need to produce more phone units.

THE INNOVATIVE LEAP – HYBRID BUSINESS MODEL

Of course, too much "virtual economy" may cause some concerns, especially when "everything as a service" comes into play. For example, we used to make software discs and sell or buy them from the stores, and it is a piece of physical product we can hold handy.

But now Software as a Service is taking over, meaning we do not or rarely produce hard discs anymore for software. Instead, people can subscribe to it as a service online on an hourly, daily, or monthly basis.

Such a "virtual economy" trend cuts off many physical resources such as business offices, retail stores, banks, and insurance agencies. There are such good reasons firms like Amazon choose to do business online as low costs, flexibility and speed, and high levels of data. It gives them unique advantages over their physical store-based competitors.

Strategically, there is no need to over-concerning the virtual economy, which is more supplementing than replacing things in our lives or say they make our ways of living more hybrid. We have more options in life: you can go to a cinema for a movie or rent and watch it from Netflix at home. You can eat out in a restaurant, or order dinner online and deliver it home.

The rule is that anything like small-size merchandise or commodity would become more eCommerce oriented. While anything large scale, such as a house, a car, or a museum, which stresses more personal customer experiences, would still need physical engagement and transactions.

Sure, we could order food over the phone a long time ago. Still, online experiences can now handle food viewing, ordering, payment, and delivery schedule simultaneously and quickly without human intervention. Such an automated process creates more business and enhances better customer experiences.

On the other hand, the risks and downside of eCommerce include online fraud, security breaches, payment safety, privacy violation, etc. We use a separate section to address such hot issues (refer to Chapter 12).

NEW MINDSET FROM ECOMMERCE

Technical Outlook – Simulating Physical Stores

E-business, in general, can perform on two leading platforms: online storefronts (DTC eCommerce (direct-to-consumer eCommerce; also referred to as D2C eCommerce) and online marketplaces. The right platform choice depends on the nature and needs of the business and audience. Either option has several benefits of its own.

Rather than involving third parties such as wholesalers, distributors, and large online marketplaces, DTC provides firms with a direct sales channel to their end-users and the flexibility to create an online store according to its needs, with customized categories and features appealing to its business model. For example, a seafood farm can sell its products online, where clients place direct orders.

Online marketplaces, on the other hand, are websites that act like business brokers. It facilitates the transactions of goods and services between vendors and customers. Examples of online marketplaces include Amazon, eBay, Etsy, Fiverr, and Upwork. On the consumers' end, more and more web traffic is generated by smartphones and tablets, driving eCommerce sales. Customers receive real-time updates on new product launches, exclusive deals, and promotional schemes. The digital one-touch purchase option has caused exponential growth in eCommerce.

In recent years, eCommerce has been leaping to the next level, empowered by three major new technologies: virtual reality, big data, and AI/ML. Virtual reality helps more clients to do online trials and design products, such as trying on clothing, touring a house, and taking online classes; big data allows the companies to understand client's needs better and purchasing patterns; AI/ML can quickly respond to and target client's needs, correct issues when they pop up and avoid human errors.

Business Outlook – Reaching Customers Directly

Online selling and purchasing offer revolutionary benefits to both sellers and buyers, which are also the reasons for the rapid growth of eCommerce. With eCommerce as the primary use, business development can be better achieved. The business relationship can be boosted with direct communication between a company and the customer.

eCommerce makes purchasing faster and easier. People can now buy stuff without the hassle of crowds, traffic, and awkward social interactions. In this fast-paced world, eCommerce offers the value proposition of a full shopping experience during the only free time available to millennials, either long commutes to work or leisure time when working from home.

Thus, eCommerce has never been more critical than ever. Understanding the latest trends in retailing, consumer expectations, design trends, and technology is essential to unlocking the potential of your ability to attract new customers and build brand loyalty.

Application Outlook – Tangible Goods with Intangible Services

eCommerce is powered by mobile business, online transaction processing, electronic data interchange (EDI), electronic funds transfer, supply chain management, Internet marketing, inventory management systems, and automated data collection systems.

eCommerce can happen via various applications, such as Email, online catalogs, shopping carts, EDI, file transfer protocol, web services, and mobile devices. The only exception is that the last-mile delivery remains tangible, which means the movement of goods through to the customer. Challenges in this area include the increasingly demanding requirements of faster deliveries and tighter delivery windows. Companies now may include using the latest technology (e.g., autonomous vehicles/drones) while managing the complexity and cost of moving goods.

An eCommerce model provides real-time data, analytics, and intelligence about products and customers from a business enhancement perspective. Such data cover how consumers interact with the site, what products interest them, and what they leave in their cart, and how much was the average purchase?, etc. Such valuable statistics allows businesses to make adjustments to meet customers' needs.

INSIGHTFUL PRACTICE

The Amazon.com B2C Story

When we talk about eCommerce, many people would quickly think about Amazon.com. True, Amazon is a very successful model of eCommerce. It started by selling books online, then it began to sell all kinds of merchandise online, including cars.

How did Jeff Bezos, the owner of Amazon.com, start the business? Jeff Bezos quit his job at an affluent Wall Street investment bank in 1994 and moved to Seattle, Washington. He opened a virtual bookstore there by working from his garage with only a handful of employees. Bezos also began developing his software for the site, which he called Amazon.com. It sold its first book in 1995.

Here is how Amazon works: businesses send their products to the fulfillment center, and the items are cataloged into Amazon's system. When things are sold, Amazon fulfills the orders and ships the products to the purchasing customers. Businesses can get transparency and track fulfillment through the Amazon back office system.

Now Amazon is the world's largest eCommerce company with consistent annual revenue growth year-over-year. The success of the eCommerce giant may be attributed to its omnichannel initiatives and diverse product offerings. Even the COVID-19 pandemic that boosted people's online activities also benefited Amazon's GMV – Gross Merchandise Value growth in 2020.

According to Statista figures (https://www.statista.com/), about 263 million American consumers shop online, some 80% of the population today. This number is projected to reach 291.2 million by 2025. The most popular eCommerce categories in the US include fashion, media, and electronics.

The Amazing Digital Story of Uber

Uber's business model is a customer-facing platform business model that connects drivers and riders and makes it easy for two sides to transact. Uber makes money from the platform's gross bookings. The business was successful because it found some particular needs and gaps that were not being fulfilled by traditional taxis or car services.

Today, 93 million customers ride or use the Uber platform. In total, 3.5 million drivers serve the growing user base. In 2022, Uber achieved Gross Bookings of $30.7 billion from its ridesharing business (source: https://backlinko.com/uber-users).

The secret weapon behind Uber's success is software and digital technologies, from GPS navigation to online payment and digital marketing. By digitizing the ride-hailing and payment processes, Uber increases customer convenience, which in turn adds quality and boosts satisfaction over that of traditional taxi-cab services.

Uber uses big data systems as a foundation for its technologies. Every ride request made by a passenger generates a lot of data. The app receives data about the passenger, location, credit card information, taxi drivers nearby, their names and car details, and the ride cost and length.

For instance, Uber uses both internal and external data to estimate fares automatically, together with street traffic data, GPS data, and its own algorithms that make alterations based on the time of the journey. It also analyzes external data like public transport routes to plan various services.

In many domains, Uber also uses AI techniques like convolutional neural networks (CNN; refer to Chapter 17). The AI can potentially forecast rider demand, pick-up and drop-off estimated time of arrival (ETAs), and hardware capacity planning requirements, among other variables that drive Uber's operations.

In a nutshell, Uber makes a disruptive technology model. It is a successful innovation that creates a new market and values, and quickly displaces established market-leading firms, products, and alliances.

Autonomous Last-Mile Delivery

Delivery is the critical phase of an eCommerce operation cycle. Every delivery faces the so-called "last mile problem" that refers to inefficiencies in last-mile delivery – primarily the need for multiple stops with low drop sizes. Last-mile delivery intends to get the parcel to the customer quickly; last-mile is considered the key to customer satisfaction.

The last mile is also the most expensive and time-consuming delivery process. It often accounts for 53% of the total transportation cost. UPS and FedEx, for example, have dominated the US logistics industry – particularly the last mile of delivery. Based on FedEx, more than 95% of all eCommerce orders in the United States are delivered by itself, UPS, or the US Postal Service (USPS).

The innovation can come from autonomous last-mile delivery, which means the parcel is delivered to end-users at their doorsteps without human intervention. Such innovations can be achieved through drones, autonomous vehicles, and robots.

Automation in the last mile intends to reduce costs, decrease manual efforts, and boost efficiency. This is why automation focuses on the most costly and time-consuming phase of the delivery process. To date, researchers have built autonomous delivery robots in experiments. They can climb stairs and hit speeds of 12.4 miles per hour. The robots are more sustainable and will help lower transportation costs.

Using Domino's Pizza as an example, customers who prepay for a pizza order on dominos.com from the participating store location now have delivery options. They can either use traditional delivery or have their pizza delivered by an autonomous vehicle called Nuro's R2 robot. As the first completely self-driving, occupant-less on-road vehicle, Nuro's R2 robot has earned regulatory approval from the US Department of Transportation.

This shows eCommerce still needs to overcome the last mile physical delivery hurdle so as to enhance their business.

Chapter 27

From Physical Banking to FinTech – Finance in the Cyber Age

This is a concept illustration of online payment and banking.

Do you still go to a physical or brick-mortar bank or an insurance agency office on the street for the financial services you need nowadays? Maybe not that much anymore. How often do you still use cash in the business and personal transactions? Maybe quite rarely. Instead, you can use all these services online or digitally thanks to the so-called FinTech or technologies for finance. Why Fintech? Because it is digital, it is more cost, time, and business effective for service providers and customers. It is a win–win.

More accurately, Fintech should be called digital technologies used in the financial service sector, including banking, stock trading, accounting, investment management, insurance, and payroll management. The new tech's goal is to improve and automate the delivery and apps of financial services.

As already described in this book, there are many digital technologies and applications in development, and why do we want to highlight Fintech here in this section? Because FinTech makes a typical and full-cycle case of digitalization in terms of technologies, systems, big data, securities, customer service, and experience.

DOI: 10.1201/9781003305187-31

Compared to other service industries such as retail, healthcare, education, and manufacturing, financial service is more critical, frequently, and widely used by nearly all adults, families, businesses, and institutions in society because financial services directly handle our money and fortunes, such as online banking, investment, insurance, billing, security, payroll, and payment management.

When we say "payment online," it means financial services are becoming paperless and digitally automated. For instance, we now use online payment systems such as PayPal, Apple Pay, Transferwise, or Payoneer.

THE INNOVATIVE LEAP – MONEY GOING DIGITAL

FinTech started to gain popularity in the 1990s thanks to the development of the Internet and eCommerce business models. This makes a turning point for people to view money and financial institutions differently.

Due to its giant scale, the digital world produces tons of data daily. This allows financial organizations to improve their business models. FinTech comes into being, leveraging AI and predictive analysis tools to detect fraud and analyze financial trends. Therefore, AI apprehends financial institutions' performance, creates insights, and forecasts algorithms.

The values and benefits brought about by FinTech are evident and enormous: it leverages technology to automate tasks and saves costs on employing people to do the work. Fintech enables organizations to utilize specialized software and algorithms used on computers and smartphones and therefore saves money by not having physical branches to service its customers.

Overall, Fintech companies usually have relatively low overhead costs, which allows them to pass the savings on to the customers. Fintech innovation around digital payments reduces costs and expands access for new customers – individuals and merchants – to payment means.

NEW MINDSET FROM FINTECH

Technical Outlook – Digital Transactions

Fintech refers to software and other modern digital technologies that provide automated and improved financial services. In contrast to traditional banks, FinTech operates flexibly and fast when implementing new services based on changing demands.

The technologies that underpin fintech business models vary considerably. They include blockchain technology, artificial intelligence (AI), machine learning (ML), big data functions such as robotic processing automation (RPA), data networking and the Internet, cybersecurity, public cloud, and mobile payment.

As the big next step for Fintech, AI-cloud integration gets trendy: AI-cloud platform applications proliferate in fields like image and audio search and processing. Deep learning will continue to apply to financial services for users via cloud platforms. One definite future of Fintech is that cash payments are obsolete and digital payments are the new face of payments.

Business Outlook – Beyond Physical Agencies

Fintech makes the most cost-effective option for consumers and businesses nowadays. It saves money because there are no hidden fees like in traditional businesses. In addition, it allows for integrating both physical and digital payment methods into a unified user experience.

Fintech today covers a host of financial activities. Starting from money transfers, applying for credit, and raising money for a business, it stretches to check deposits using a smartphone and managing investments. As the society follows a digital transformation, many companies and consumers are adopting digital payments, mobile banking, marketplace lending, money transfers, and many other fintech services. This trend is driving demand and growth of the fintech market.

For instance, with the Fintech setup, commercial institutions that face businesses or entrepreneur bank accounts now can provide novel approaches to personal loans, providing more people with lending options and faster, easier accessibility and experiences than traditional financial institutions.

Application Outlook – Easy, Fast, and Secure Payment

There are four major categories of users for Fintech adoption: (1) between the banks, (2) for banks' enterprise clients, (3) for small businesses, and (4) for consumers. Fintech apps, again, have changed the way businesses function. Electronic transactions have become much easier and faster for financial transactions. Fintech is answering well to the need for a more effective financial system.

Within Fintech, we have identified 12 distinct subcategories and an additional "other" subcategory. The 12 distinct subcategories are (in alphabetical order): (1) accounting and expense management, (2) blockchain and crypto, (3) capital markets, (4) digital banking and financial infrastructure, (5) HR, payroll, and benefits, (6) insurance, (7) lending, (8) payments, billing, and money transfer, (9) personal finance, (10) real estate, (11) regulatory tech, and (12) wealth management.

Of course, some side effects of Fintech must be monitored and managed closely. The risks posed by Fintech to consumers may include loss of privacy, compromised data security, and the rising dangers of fraud and scams. Also, some FinTech practices may lead to unfair and discriminatory uses of data and data analytics, non-transparent data use to both consumers and regulators, and harmful manipulation of consumers.

INSIGHTFUL PRACTICE

Capital One – A Software Powerhouse

Capital One is a primary financial institution and Fortune 500 company. As one of the primary financial institutions in the United States, Capital One offers a wide variety of financial products in retail and corporate banking. It has a strong base of more than 40,000 employees.

The company specializes their work in banking, and savings accounts, credit cards, auto loans. It is HQed in McLean, Virginia, with operations and market primarily in the United States. In 2021, 60% of the company's $25.77 billion revenues were from credit cards, 23% were from consumer banking, 8% were from commercial banking, and 3% were from others (source: https://www.capitalone.com/).

Back in 2010, Capital One, like most other banks, was still outsourcing most of its software-development activity and did not consider technology as genuinely strategic. Today, the company is a software powerhouse with a technology staff of almost 11,000, 85% of whom are software engineers. Capital One's objective is to change banking for good by bringing simplicity to banking.

Its AI and ML capabilities are at the forefront of what is possible in banking today. Many of the firm's AI and ML technologies, such as Eno Virtual Credit Card Numbers, CreditWise, and Auto Navigator, are firsts in the industry.

Capital One online banking lets customers choose when, where, and how to bank. Customers can access their accounts online or through the Capital One Mobile app. With built-in security and 24/7 access, customers' money is always at their fingertips. Online banking allows customers to manage their money anytime, almost anywhere.

Meanwhile, Capital One uses a hybrid approach to meet all customer needs. At the bank's full-service branches, you can get a cashier's check, make deposits, cash checks, and open a checking or savings account. Customers can get the help of friendly Ambassadors onsite. With the bank's network of ATMs all over the country now powered by IoT, it is simple to check your balance and get cash with no added fees as a Capital One 360 Checking account or Performance Savings customer.

AI and Modern Online Banking

AI in FinTech is adopted for many tasks: lending decision-making, credit risk assessment, insurance, customer support, fraud detection, wealth management, and more. FinTech companies adopt AI for enhanced efficiency, higher precision levels, and fast query resolution. Today, banks realize that data science powered by FinTech can significantly speed up these decisions with accurate and predictive analytics.

Banks can apply AI to transform the customer experience because AI enables frictionless, 24/7 customer service interactions. Meanwhile, AI in banking applications can go beyond retail banking services. The banking value chain, such as back and middle offices of investment banking and all other financial services, could also benefit from AI.

The banking sector extensively uses AI and ML to automate processes and makes them more accessible. A few major use cases where these emerging technologies are used are for fraud detection, theft prevention and recovery, and security breaches around sensitive information and cash.

AI-powered FinTech systems can appraise customer credit histories more accurately, which helps avoid human errors and routine default levels. Mobile banking apps easily track financial transactions and analyze user data. AI tools help banks anticipate the risks associated with issuing loans, such as customer insolvency or the threat of fraud.

With its power to predict future scenarios by analyzing past behaviors, AI helps banks predict future outcomes and trends. This allows banks to identify fraud, detect anti-money laundering patterns, and make customer recommendations.

One may have a question though: considering Fintech is already so powerful and predictive, then why banks like SVB (Silicon Valley Bank) still fell bankrupt all of a sudden in March 2023 and nearly caused a broad financial crisis? In this case, the monitoring Fintech system actually did send out risk alarms quite a few rounds to SVB executives, but failed to get their attention and corrective actions to be taken on time. This sets a good example that digital power alone still not good enough in such strategic cases. It is the hybid human and digital power that can help make real differences in business and life.

Pros and Cons of BitCoins

One of the hot cases of Fintech in action is cryptocurrency, such as Bitcoin. It is forging a new segment of Fintech called CryptoTech Industry. A cryptocurrency is a medium of currency exchange, such as the US dollar. But the difference is that it is digital and uses encryption techniques to securely control the creation of monetary units and verify the transfer of funds. Bitcoin (BTC) is the name of the best-known cryptocurrency for which blockchain technology was invented. Digital currency tries to provide an alternative payment system to what we are used to now and is created as a method for people to send money online.

Cryptocurrency aims to aid the transaction of goods and services in a safe and secure online environment without central control but be used like traditional currencies. BTC likes to be anonymous and the process would need very little or no government and middleman interference. BTC, therefore, can lower the cost of influencers and reduce the time of transactions. It uses a peer-to-peer Internet network to confirm purchases directly between users.

With the inception of BTC, the government can lose control over the currency system due to BTC's decentralization. BTC's underlying technology does not need any central authority for any transaction, and the government cannot regulate the monetary policy and loses its power. Thus, some economies do not like BTC because the so called central control can have twofold meanings: it can mean financial monopoly, but it can also mean better security measures against fraud and scams.

Some countries thus have placed limitations on using BTC, and banks there may ban customers from cryptocurrency transactions. Even worse, some other countries and governments have banned the use of BTC and cryptocurrencies outrights with heavy penalties.

Although BTC uses secure cryptography, you could argue that it is not a safe investment because of its volatility. With no regulatory body and an international 24/7 market, a BTC worth USD$60,000 one day can be worth USD$30,000 just a few days later. Though there have been some periods of stability, these never last long.

Nowadays, items such as insurance, consumer staples, luxury watches, and event tickets can be purchased via cryptos. If one wants to buy things with cryptos, start with getting a debit card. Available from major crypto exchanges and other providers, the debit cards permit the holder to withdraw cash from participating ATMs.

This shows Bitcoin is a digital currency that can be used for purchasing online.

Part V

Layer 5 – Wisdom in Digital Space – Equilibrium, Leadership, and Culture

Layer 5 is the top level of the Cognitive Model of Digital Transformation introduced by this book and the ultimate objective the digital mindset leaps should reach. At this layer, your mindset, values, and behaviors will go beyond technologies and their applications. You will care more about and better manage the new relationship between humans and machines in the digital era. You do not want to fall behind the technologies, meanwhile do not want to get overwhelmed by technologies either. Only those who can take the lead and manage this new human-tech relationship can achieve the greatest. Once at this level, you've obtained an optimal, innovative, and sustainable digital mindset.

> It is not the strongest of the species that survives, nor the most intelligent that survives. It is the one that is the most adaptable to change.
>
> – Charles Darwin, English Naturalist, Geologist, and Biologist

> We talk about automating operations, about people, and about new business models. Wrapped inside those topics are data analytics, technologies, and software – all of which are enablers not drivers. In the center of it all is leadership and culture. Understanding what digital means to your company – whether you're a financial, agricultural, pharmaceutical, or retain institution – is essential.
>
> – Jim Swanson, Senior Vice President/CIO and Head of Digital Transformation, Bayer Crop Science

DOI: 10.1201/9781003305187-32

Chapter 28

From Data Protection to Trusted Cloud – Digital Rights and Safety

This demonstrates online privacy is important amid our keyboards today.

As we have generated and collected more and more digital data for processing, storage, and services, the data also are facing the severe challenge of protecting clients' and civilians' privacy and confidentiality. While data are used to serve customers and markets better, they are also very vulnerable to malicious attempts or abuses.

Amid the challenges, the top three parties most frequently involved are companies, governments, and hackers. The beneficiaries or victims are, of course, clients or consumers. Nowadays, we often see such breaking news that the online databases or storage of some banks or government agencies are breached and compromised by hackers, or some governments forced technology firms to hand over clients' data for national security investigation purposes, and so on.

Why is data privacy so important? Because data can be person and organization specific. When private data get into the wrong and malicious

DOI: 10.1201/9781003305187-33

hands, bad things can happen. A data breach happening to a government agency can, for example, put top-secret data in the hands of an enemy state. A breach in a corporation database can put proprietary and confidential data in the hands of a competitor or rival.

Personal data protection is not just about protecting people's data. It is to protect the fundamental rights and freedoms of people while safeguarding personal data and helping ensure that persons' rights and freedoms are not violated. Examples include the children's Online Privacy Protection Act (COPPA) which gives parents control over what information online for their kids.

When companies risk clients' data exposure for commercial interests, customer privacy and confidentiality are in jeopardy. When governments force data handouts in the name of data sovereignty, national security, taxation audit, etc., businesses and clients may fall victims. When hackers attack the system and steal data, all others may fall victims.

THE INNOVATIVE LEAP – DIGITAL DATA PRIVACY POLICIES AND PRACTICE

Hence when talking about clients and consumer data protection, there are two major types of measures to take: one is technical tightening up (refer to Chapter 12), and another is policy and regulation compliance; both types are equally important. After all, data security is handled by humans and also is about humans. In particular, in each different and complicated situation, what proper measures can be taken to protect the clients' data effectively? Big technology firms together now try to address this critical question.

In September 2021, Amazon, Google, and Microsoft joined other tech heavyweights such as IBM, Salesforce/Slack, Atlassian, SAP, and Cisco in agreeing to the Trusted Cloud Principles (source: https://trustedcloudprinciples. com/principles/), a broad set of principles that say that these companies are committed to the privacy protection and security of their customers' data in all jurisdictions through policy and technology.

The Trusted Cloud Principles were developed to address the industry's concern that certain regulations and proposals will make it hard to do business and may force companies to hand over customer data without being able to notify customers.

By nature, the Trusted Cloud initiative is to address the balance between customer and government interests in case of any conflicts. The companies are committed to working with governments to ensure digital connectivity among nations and promote public safety while protecting cloud privacy and data security.

NEW MINDSET FROM DATA PRIVACY PROTECTION

Technical Outlook – Client's Consent First

Trusted Cloud means that all client information held via any media such as on paper, computer, visually or audio recorded, or in the memory of the professional, normally without the client's consent, must not be disclosed.

Confidentiality means non-disclosure. Namely, without the client's express consent, personal information shared with an attorney, physician, therapist, or other professional individuals generally cannot be divulged to third parties. Confidentiality is regarded as an ethical duty, while privacy is a right in the common law. This means that violation of confidentiality can lead to legal lawsuits against privacy.

Lawfulness refers to any processing of personal data that should be lawful and fair. Fairness and transparency mean it should be transparent to individuals and consumers that their personal data are collected, used, and consulted. If there are exceptions, the client is entitled to know to what extent the personal data will be processed.

Personal Identifiable Information (PII), or personal ID, is one of the biggest concerning areas in data privacy. On the one hand, the veracity and volume of such data in our technology-driven world are unprecedented; on the other hand, it becomes challenging to handle possibly even billions of data records properly and legally.

Business Outlook – Follow the Data Laws and Policies

Organizations that collect data from clients must respect and follow the laws within their IT security framework, which consists of a series of documented policies and processes for implementing and maintaining information security controls.

The policies serve as a blueprint or plan for managing risk, reducing vulnerabilities, and ensuring that the subjects can access and control their data. The policies and processes also determine who may access it and whether it is shared or sold. Such a security framework governs as a system of standards, guidelines, and best practices, and it prioritizes a flexible, repeatable, and cost-effective approach to promote the protection and resilience of your business.

For instance, data protection in the framework manages risks around data collecting, processing, storing, and reporting. It is a systematic approach to ensure that data is maintained securely, preventing important data from compromise, corruption, or loss. It provides the capability to restore the data to a functional mode should something malicious happen to render the data inaccessible or unusable.

Application Outlook – Data Security Solutions

Traditional data security solutions include encryption, data loss prevention (DLP) technology, backup and recovery solutions, identity and access management technology, and more. But frequently overlooked when designing a data security framework is the role the application security can play in protecting data.

Application security features include authentication, authorization, encryption, logging, and application security testing. Developers can also code applications to reduce security vulnerabilities. The top three most common risks for application security are broken access control, cryptographic failures, and injection. An injection attack, for instance, is a risk of vulnerability where an attacker relays malicious code through an application to another system that may impact servers and clients.

Hence, we need to implement all app security measures and solutions to minimize the chances that malicious hackers can breach data, systems, or applications, and that is the ultimate goal of application security – prevent attackers from accessing, modifying, or deleting sensitive or proprietary data.

INSIGHTFUL PRACTICE

Apple vs. FBI in iPhone Encryption

In 2016, this story went viral: the Federal Bureau of Investigation (FBI) wanted Apple Inc. to create and electronically sign a new software agreement. This would enable the FBI to unlock an iPhone 5C it recovered from a crime scene. The suspect was involved in a December 2015 terrorist attack in San Bernardino, California, when 14 people were killed and 22 injured.

Apple's privacy protections had stalled the FBI's investigation for months. Finally, the US Justice Department said it had accessed a terrorism suspect's iPhone without help from the company. Apple remains consistently in the position that it will not re-engineer its phones for law enforcement.

At the time, we noted that there were essentially two possibilities: If these were phones with known vulnerabilities, the FBI could pay as little as USD $15,000 to have an external firm break the encryption; or if they were newer models, then even Apple would likely be unable to help.

Eventually, Azimuth Security, an infosec company based in Sydney, Australia, assisted the FBI in the hacking solution, and two Azimuth hackers succeeded in breaking into the iPhone 5C, according to a *Washington Post* report (https://www.washingtonpost.com/technology/2021/04/14/azimuth-san-bernardino-apple-iphone-fbi/).

Usually, end-to-end encryption provides the highest level of data security. Because Apple's operating system is a closed system. Apple doesn't release or disclose its source code to app developers. Hence the owners of iPhones and iPads cannot modify the code on their phones themselves. The pro part is that this makes it more difficult for hackers to breach your iOS-powered devices.

On each apple device, the data that the user stores in iCloud are associated with the user's Apple ID. The data are protected with a key made from information unique to that device, combined with the user's passcode, which only the user knows. In general, this provides robust security to the phone user.

Digital Data Sovereignty

From a regulatory angle, data sovereignty generally refers to government measures to prevent citizens' data from falling into the wrong hands. The measures will restrict how businesses can transfer personal information beyond their borders. Legally, data sovereignty means that digital data is subject to the laws of the country where it is located. Data sovereignty is about protecting data and ensuring it remains under its owner's control and within the boundaries of the country involved.

The rights for digital residency and security are the right to be informed, the right to rectification, the right of access, the right to object, the right to erasure, or to be forgotten. Such rights directly restrict the rights for data portability and processing and the rights concerned about automated decision-making and profiling.

Many countries such as Germany, France, and Russia have established laws for data sovereignty. Such laws require citizens' data to be stored and processed on physical servers within the country's borders. To date, more than 100 countries now require their citizen data to be stored in servers physically located inside their borders.

Data sovereignty helps cope with complicated digital asset issues and regulate information flow online. In theory, an international set of regulations would save many of the requirements and helps to decide data sovereignty for a particular data set.

Digital sovereignty is a key concept in the Internet age, considering and measuring how data and digital assets are treated. It requires that users have sovereignty over their own digital data, either individually or toward nations.

The Edward Snowden Story

Edward Joseph Snowden is an American former computer intelligence consultant. In 2013, he made public highly classified information from the National Security Agency (NSA), where he was an employee and subcontractor.

In early 2013, Edward Snowden, a computer specialist and former CIA systems administrator, made confidential government documents to the public press about the practice of government surveillance programs. In their conclusion, many legal experts and the US government insist that Snowden's actions violated the US Law – Espionage Act of 1917. The law identified the leak of state secrets, such as what Snowden did, as an act of treason.

In June 2013, the US Department of Justice (DoJ) unsealed charges against Snowden for two counts of crimes: violation of the Espionage Act of 1917 and theft of government property. Following the charges, the Department of State revoked Snowden's passport.

Yet Snowden argued that he was morally obliged to act even though he broke the law. He tried to justify his "whistleblowing" and stated it was his duty to inform the public, "which is done in their name and sometimes against them." Mr. Snowden believes the government's violation of citizens' privacy had to be made known to the public regardless of legality.

Edward Snowden's escape to Russia is part of the aftermath of the global surveillance disclosure incident. Snowden flew to Hong Kong first and then to Moscow. The Russian government allowed him to travel freely within Russia and recently granted him Russian citizenship.

Many agreed with Snowden and defended his actions as ethical, while may not be 100% legal, arguing that he acted from a perspective of doing public good. They feel Snowden may have violated a secrecy agreement. But the agreement is not a loyalty oath but a contract. It is a less important deal than the social contract a democratic society has with its citizenry. Some others argued that even if Snowden was legally guilty, he was not ethically guilty because the law itself was unjust and unconstitutional. The Snowden controversy may go on.

This is a sign of online defense shield as a symbol for cybersecurity.

Chapter 29

From Digital Addiction to Mindfulness – Digital Era Equilibrium

This shows IQ+EQ together makes a winning formula for the digital age, and IQ alone will not suffice.

Many new technologies may address a certain professional group and rarely get very personal to the general business and public. But digital transformation (DX), as defined by the 4th Industrial Revolution (4IR), is unprecedented and tends to get immersed with all the businesses, organizations, institutions, and each one of us in the 21st century. We are entering a new era and social environment where we cannot live well or function right without digitalization.

So, what is the best way or approach for us to stay up with or even stand in the front of the DX trend? On the individual level, you may have heard advice such as setting up some road map and goals for yourself, starting small, and keeping up with best practices and learning.

This book, in particular, advocates starting a DX journey with mindset adaption, namely, being open, mindful, and immersive in your digital life and practice. Being open means adapting to new technologies and the changes they bring about. We all enjoy the benefits of digitalization, but some are concerned with and even afraid of the learning curves. Hence,

DOI: 10.1201/9781003305187-34

keeping an open mind to new things, digital apps, and technology horizon is the critical attitude to take amid all the rapid changes.

Be bold to try out new things and innovative solutions. The first impression of digital technologies and applications might be that they look complicated to understand, operate, and have so many functionalities to learn about. Still, once you take the first step of doing or using digital devices and apps, you will gradually learn and grasp the new solutions. For instance, it is time to switch to digital banking (instead of still going to a bank branch in the street) to make use of those fintech apps and services that banks offer on your mobile phone, tablet, or computer nowadays.

THE INNOVATIVE LEAP – ACHIEVE YOUR EQUILIBRIUM OF DIGITAL LIFE

Meanwhile, try to stay mindful of all the changes, namely, keep a calm and clear head in the new waves of data explosion, online experiences, and security and privacy challenges. Keep the balance between the new technology world and the world of nature we live in. Learn how to choose the optimal path in handling things in the digital era and be aware of the pros, cons, benefits, and risks.

One best practice is to avoid getting addicted to those tech-backed behaviors and habits, such as a person every one moment holding and checking on smartphones every day, young kids getting lost in online gaming plays, overdosing on online social media, and networking with big exposure to health, security, and privacy risks.

Another best practice is when doing online activities like purchasing and banking, always be mindful of security in this process. It is essential to use your trusted browser and via secure websites only. Remember for example, that the website address you are browsing must start with HTTPS. Websites with addresses beginning with HTTP only are unsure, and information over such websites can be intercepted by hackers easily.

Also, it is critical to pay attention to other key items such as passwords, the apps we use, the safety of our mobile phones, and bank card. Signing up for alert messages on payment transactions is beneficial to help you monitor all your debit/credit-card activity. You will be notified of any suspicious activity immediately to contact your bank.

Meanwhile, have fun and enjy the benefits in your digital practice, like making your home a smart home, using ChatGPT to aid your work and study, implemneting automation in your business operation, and do not wait until close to the end of the journey or the completion of your DX curve. Feel free to celebrate your milestones. After all, DX is a journey for each of us to live through and is supposed to bring better lives and good time to us in general. Hence it should be fun and awarding.

NEW MINDSET FROM MINDFULNESS IN THE DIGITAL ERA

Technical Outlook – Be Wary of Technology Dependence

Overwhelmed by the magic and convenience created by technology, many people have become too dependent on technology. As more our everyday life activities are managed through technological gadgets and applications, people now rely on smartphones to check the time and constant communication throughout the day. While technology has made life easier, safer, longer, and more enjoyable in numerous ways, it has its dark side too. Hence it is wise that we find a balance in how to use and cope with technologies.

Positive effects of technology include less expense, better efficiency, more communication channels, an increase in productivity, etc. Negative effects of technology include social isolation, job loss, adverse health effects, security breaches and scams, etc. For instance, relying on computers has both positive and negative effects. It is true computers have proved being much more reliable than humans. Therefore, some consider that computers never make mistakes, they never take rest, and always are faster than humans.

But we are paying the price for our digital life because of digitalization's dependence and virtual lifestyle. And some studies have found that excessive dependence on cell phones and the Internet is akin to addiction. This poses a more serious challenge than in the past when the central concern was that technology detracts from interpersonal relationships and social norms.

Business Outlook – The Only thing that Does not Change is the Change

An effective DX is good at bringing about and coping with changes. Since DX integrates and applies digital technology to all business areas, it is fundamentally changing how we operate and deliver value to customers.

With DX, companies can take a step back and revisit everything they do, from internal systems to customer services, both online and in person. It is also about cultural and mindset changes that require organizations to continually challenge the status quo, conduct experiments, and get comfortable with testing and failure.

Grow a digital culture that describes how ICT technology shapes how we interact as humans, showing the organization new and digital ways we behave, think, and communicate within society. The culture change will foster positive adjustments whenever you find things conflicting with the track of DX or clashing with your digital project goals.

Application Outlook – Reserved with Digital Apps and Devices

Try to be binary in digital technology use, namely, know when to be on and off. Mindful technology use is also about minimalism and away from addictiveness. That does not mean you need to restrict your use of technology or use as little as possible. Instead, it means avoiding wasted technology use and not using more apps than you can reasonably handle, and focusing on the technology tools that are most beneficial to you.

Modern mindfulness indicates developing wisdom in your daily routine and digital device use habits. Now you pay attention to what is important by eliminating digital distractions and interruptions as much as possible. For example, grow a good habit of pausing and paying attention to nothing every day. It helps you to build resilience to cope with the fast changing technological circumstances and the negative impact it may have on our physical and mental well-being.

Other good mindful habits include: only accessing content and enabling needed notifications; avoiding content you know will trigger negative reactions; keeping a distance between you and the screen; trying to stand up while working instead of long time sitting; taking opportunities for real human contact; and keeping some awareness of your breath to become mindful and retain mindfulness.

INSIGHTFUL PRACTICE

Negative Sides of Digitalization

DX has changed a lot in our daily lives, such as personal relations, business, communication, and how everyone uses their time. The impact of technology is always associated with achieving greater freedom. Still, it has also pushed us to levels of distraction and dependence that have changed how to engage with the world.

Information and communication technology (ICT) also has dark side effects such as spam, malware, hacking, attack,and violation of digital property rights. ICT also has non-technology-centric dark side effects like online theft, cyberbullying, and online addictiveness.

For enterprises, because your business benefits from technology, your company also becomes dependent on it. When the inevitable glitches, bugs, and power failures come, you may be unable to accomplish simple business tasks. For example, credit-card processing operation glitches will create major headaches for both stores and shoppers.

Other negative digital examples include potential major privacy and security breaches and the proliferation of fake news. Some may become criminal in nature with harmful actions such as profanity, cyberbullying, hate speeches, , online predation, sextortion, and gaming addiction, plus the negative effects of the Internet on social relationships and social cohesion.

Based on various research in recent decades, social media and mobile devices may result in psychological and physical issues, such as eyestrain and distraction from key tasks. Even worse, online activities may contribute to more serious health conditions like depression. The overuse of technology is pretty much like drug overdosing and may have a more negative impact on developing children and teenagers.

It is easy to become addicted online. Excess electronic and digital gadgets can make children spend less time outdoors and hurt their social capabilities. Poor concentration in studies and losing interest in day-to-day activities are also major concerns.

Research shows that youngsters who spend too much time on social media can suffer from the following issues: poor sleep, eye fatigue, negative body image, depression, anxiety, cyberbullying, and more. In summary, excessive gadgets use can result in poor health, a chaotic lifestyle, and bad eating habits.

Mindfulness and Equilibrium in the Digital Era

Mindfulness and its practice originated from ancient eastern culture and Buddhist philosophy back around 2500 years. This ancient wisdom of mindfulness was introduced to the western world around the 1980s.

This is a demonstration on what digital equilibrium is about, meaning one got to know how to keep a good balance between different terminals in the digital era and stay mindful.

Mindfulness comes from meditation, for example, the contemplation practiced during yoga exercises. It is the method and state where one attains samadhi – a state of meditative consciousness. This is where the mind becomes very still and merges with the object of attention.

Mindfulness cultivates resilience in our nervous system, which is a magnificent self-organizing structure. We must practice and re-discover the wisdom and strength within this digital and technological age. This will help us keep in good touch with the human and reality moment – when we are present to listen to each other.

Applying mindful technology use and digital mindfulness involves developing the structure of your daily routine. For instance, smartphone use habits are a good starting point. We should learn to pay attention to what is important by eliminating digital distractions and interruptions as much as possible. Being digitally mindful is taking a moment to pause and pay attention to your thoughts and feelings. It helps build resilience to cope with your unusual circumstances and their impact on your physical and mental well-being.

What are the seven principles of mindfulness?

Non-judging or labeling	Be an impartial practitioner of your own experience.
Developing patience or no rush	Patience is the wisdom that demonstrates that we accept the fact as they come and go.
Keeping a beginner's mind	Being open and curious allows us to be receptive to new.
Building trust	Develop essential confidence and trust in yourself and your feelings
Living the moment	By habit, bring open, discerning, and accepting attention to everything you do
Using acceptance	Accept yourself and treat yourself nicely the way you would treat a good friend.
Letting Go	Focus on your breathing with little distractions

Mindfulness is a personality trait strongly associated with flexible responses to stimuli, increased subjective well-being, and reduced psychological and physiological symptoms. Being mindful means staying aware of your thoughts, emotions, and physical feelings. By doing so consistently, you can cultivate yourself to live in the current moment and enjoy life as it happens.

Online Gaming Without Addiction

In 2019, Entertainment Software Association (ESA) surveyed over 4000 households in the United States on online game-playing activities. It reported the following: the average age of a US gamer is 33 years old. The average gamer has been playing online for 14 years (source: https://intenta.digital/gaming/video-game-industry/).

Many successful people enjoy video games. Unfortunately, they do not always talk about it due to the stigma around video games. But this is strangely backward: video games do not hinder success; they enable it.

Gaming is considered a form of entertainment, and may make one smarter. A new study by the National Academy of Sciences (source: http://www.nasonline.org/) demonstrates that people who have played action online games, such as "Call of Duty," "Unreal Tournament 2004," etc., show a greater capacity to learn than those who played non-action games.

Meanwhile, we need to be on guard against video game addiction which is compulsive or uncontrolled video game plays. Often that leads to health problems in the person's life. Gaming addiction can be a compulsive mental health disorder that, if not treated on time and right, can cause severe damage to one's life. One typical symptom is that a video game addict may spend over 10 hours a day gaming. They usually play all the way into midnight, and many suffer from sleep deprivation.

How can gaming become an addiction? When a person engrosses themself in playing video games and starts to feel hyperarousal, their brain could associate the activity with dopamine. That means the person develops a strong drive to seek that same pleasure again. Thus, the American Academy of Pediatrics recommends screen-based entertainment for no more than 2 hours per day.

Many games require complete focus. Excessive video game playing can cause eye discomfort, focusing problems, blurry vision, and headaches. It is so easy to completely immerse in a game, distracting the person from natural breaks. Here are some ways to prevent a gaming problem: First, set time limits for play and stick to them. Second, keep phones and other gadgets out of reach at night so you will not play into the night. Third, make sure you have a balanced daily routine: do other daily activities, most importantly, including exercise. This will reduce the health risks of sitting there and playing for long periods.

Chapter 30

From Entrepreneurs to Digital Leaders – New Vision and Horizon

This is a typical leadership showcase in leading the team to navigate the unknown water and head to the destination of success.

Digital transformation (DX) refers to integrating and immersing digital technology into all businesses. It fundamentally changes how enterprises and organizations operate and deliver customer value in terms of technology, data, process, and structure. Organizational wise, DX can be regarded as the corporate-led strategy for technology implementation to reach a certain point or scope or end of the value.

As covered entirely in this book, digital technologies and applications are electronic tools, devices, systems, and resources that generate, process, and

DOI: 10.1201/9781003305187-35

store data. Examples include social media, online apps, multimedia, mobile phones, IoT, cloud computing, smart home, operative automation, etc.

As stated in the Preface of this book, some people may wonder, the 3rd Industrial Revolution (3IR) already included PCs and the Internet that are all digital; what really is the difference we are now experiencing? In a nutshell, we can say this way, while in 3IR, digital technologies remained as our tools and assistance, now in 4IR, digital technologies such as cloud, SDN, NVF, and 5G combined with AI are becoming part of us intrinsically, and integrating with our daily work operations and life activities. That is why we call this change process DX. It is a transformation.

DX calls for new types of leaders with vision and implementation power as an organization navigates through new water. Digital leadership is the strategic direction and use of an organization's digital assets to achieve business goals. Digital leaders are willing and open to exploring how ICT (information and communication technology) can be used to enable an organization to better fulfill customer needs and changing business requirements and ultimately transform the whole organization into a smart company.

THE INNOVATIVE LEAP – VISION TO LEAD DX

A transformative vision is the most important leadership skill and area in the digital era. It includes anticipating market and technology trends, making savvy business decisions, and solving tough problems in turbulent times, adopting the right technologies, making savvy.

One of the primary visions for new digital leaders to share with the organizations is on the impact of new technologies, which can reduce the workload and help add more values. As a digital leader, your role will be primarily to engage your organization's employees and foster a culture of exploration of the opportunities that technology presents. Businesses that focus on technology only as an objective are the ones that fail because the outlook is too narrow.

Successful digital leaders have the following traits: try and thrive despite uncertainty – dare to look into new horizons for the company, enthusiasm, and quest for the breakthrough of business outcomes – never be satisfied with the status quote, and want to get from good to great; exploit new digital competitive levers – stay abreast with new technologies and game changers; start, experiment, learn, and train; grow and cultivate creativity and a culture of innovation; and attract and hire top-notch digital dexterity talent.

NEW MINDSET FROM DIGITAL LEADERSHIP

Technical Outlook – Evangelizing Digitalization

A digital leader understands new technology gist and can explain technical messages well to non-technical people. These leaders transfer complex ideas from one team member to another or simplify an intricate vision for the whole firm.

Typically, tech leaders initially emerge from the creation of new technology advancements. A technological leader can make something happen either in the technology (i.e., causing technology to appear or be used productively like Steve Jobs created the smartphone) or aided by the technology (i.e., causing things to occur using productive technology like Jeff Bezos established Amazon.com).

A successful digital leader is innovative, creative, collaborative, experimental, curious, and able to network. They are forward-thinking, industry-leading, and, most importantly, can foresee a constantly changing landscape.

Business Outlook – Envisioning the Roadmap for Digital Transformation

Digital leaders will set the vision, influence people, define processes, look for continual improvement, and track impact. Although businesses must have a stable and secure infrastructure, people are undoubtedly the most important factor in digital transformation, not technology.

A digital vision should bring about a compelling picture of the transformation. It should foster awareness and education at all levels of the organization. This will empower your teams to help you successfully create the new reality you envision.

A clear roadmap would help make a sound digital strategy that is good at using technology to improve business performance. It could be creating new products or re-creating current processes. It guides an organization's direction and process to create new competitive advantages with technology and enables the tactics it will use to achieve these changes.

For instance, bringing artificial intelligence into your service organization is the power showcase of DX. You may start to implement AI-powered chatbots that answer simple customer inquiries. They also serve as a welcoming presence on your website, reducing the time customers wait to reach an agent.

Application Outlook – Hands-on with Digital Projects and Programs

A capable digital leader understands the importance of data generated by the business and the processes that support and make it happen. Digital leaders place a high value on their communication and creativity; They are willing to explore new emerging technology and digital information and help organizations find cutting-edge alternatives to outdated programs and legacy systems.

Digital leaders approach problems with high curiosity. They embrace an open mind and always encourage creativity. For instance, how can you roll out new software or device to your team without adding backlash and growing pains? Here is a approach example to successfully implementing new technology in the workplace:

Step 1	Explore technologies that will solve problems for your company
Step 2	Build an implementation team to champion the new technology once you have chosen it
Step 3	Implement the technology through a trial or pilot program to work out uncertainties and gain buy-in
Step 4	Train your employees to learn and use the new tool
Step 5	Launch, optimizing the tool to fit your needs as time goes

INSIGHTFUL PRACTICE

Providing Apple to the World – Steve Jobs

Steve Jobs was a great pioneer of the personal computer (PC) era and then a champion in the smartphone era. Jobs co-founded Apple Inc. in 1976 and later transformed the company into a world leader in telecommunications via the birth of the iPhone. Well known for his vision and genius, he oversaw the development and launch of such next-gen products as the iPod and the iPhone.

This is a portrait of Steve Jobs – the co-founder of Apple Inc. and inventor of smartphones we use today.

In 1976, when Jobs was just 21, he and his partner started Apple Computer in the Jobs' family garage. They funded their entrepreneurial venture by selling their Volkswagen bus and beloved scientific calculator.

Steve Jobs was very successful because of his innovative ability to design perfect products. This has attracted immense market demand from consumers, and Steve Jobs is the major contributor to Apple's rapid growth and development due to the innovativeness of its products and services.

Steve Jobs' leadership style was quite autocratic, and he had a sharp eye for detail and hired like-minded people to follow his lead. His creative awareness and detail orientation were driving success. Jobs was telling people always be "hungry" and willing to learn new things. Stay hungry means always being curious to learn more and achieve more.

Most people believe that his constant need for perfection drove Jobs to persevere through projects, but his biggest motivation was his desire to leave something behind that changed everything. He wanted to change the world and technology for the masses. He not only transformed the core business of personal computing but also changed the world of music, phones, tablets, digital publication, retail stores, and animated movies.

Steve Jobs' legacy and impact on the world continue today through his accomplishments in technology, innovation, and new product development. Here are the six leadership qualities of Steve Jobs:

- Invent new and simplify the existing. By record, Steve Jobs is the primary or co-inventor of 241 patents.
- Hire and develop the best. Steve Jobs believes, "The greatest people are self-managing."
- Have a clear vision
- Focus
- Customer obsession
- Deliver results

Make Shopping Happen in Online Clicks – Jeff Bezos

Jeff Bezos is an entrepreneur, eCommerce pioneer, and the founder of the eCommerce company Amazon. He is also the boss of the *Washington Post* newspaper and the space exploration company Blue Origin. Mr. Bezos' business successes have made him one of the richest people globally.

Born in 1964 in New Mexico, Bezos loved computers as a child and later got enrolled in computer science and electrical engineering at Princeton University. He was in New York City from 1990 to 1994, working on Wall Street for a hedge fund firm. One day however, when he came across the promising statistic that "online web usage was growing at 2,300 percent a year," that changed his life and inspired him to start a new business that would "make sense in the context of the Internet growth." So, he quit his lucrative job in New York city and went to

Seattle to open Amazon.com, an online bookstore that later became one of the Internet's biggest success stories.

Settling in Seattle, Washington, Bezos started up a virtual bookstore out of his garage (Both Jobs and Bezos have proved the house garage can be the venture and fortune birthplace) with a handful of employees. Soon Bezos began developing the software for the site, which he called Amazon.com. Amazon was founded on July 5, 1994, and it was a website that only sold books, and it sold its first book in 1995. The business model followed Jeff Bezos's vision for the company's explosive growth and eCommerce domination. At the start, he knew he wanted Amazon to be "an everything store."

Bezos is known for his 14 leadership principles in establishing and developing Amazon. First among them: think long term. Bezos focuses on big, long-term goals and ignores the short-term trivialities that distract others. Bezos's second principle is to be ferocious. He speaks endlessly about "customer obsession." He proudly tells job applicants that Amazon employees ("Amazonians") work long and hard.

Amazon made shopping more convenient through features like one-click ordering online, package pickup at Amazon hubs and lockers, personalized recommendations, and in-home delivery with Amazon Key. Bezos became a millionaire in 1997 after Amazon's initial public offering (IPO). He was first entered on the Forbes World's Billionaires list in 1999 with a net worth of USD$10.1 billion.

Bezos has succeeded because he was always thinking three years ahead. He believes he gets paid to make a few high-quality daily decisions, stating his goal is to make "three good decisions a day with high quality, and they should just be as high quality as I can make them."

Constantly Exploring New Horizons – Elon Musk

Elon Musk may be the most famous man in the world today. Elon Musk wears many hats: he cofounded the electronic payment firm PayPal and founded the SpaceX spacecraft company. He is also the CEO of the electric-car maker Tesla, and the globally influential social media firm Twitter.

There is a reason for Musk's success: he is a highly inspiring leader. While some of his ideas may be unfeasible, he owns the critical ability to motivate people and get them excited about his ambitions, plans, and projects. He makes people believe in his vision.

Musk was a South African native and grew up there. He then stayed in Canada before moving to the United States. He graduated from the University of Pennsylvania in physics. Musk started it off as a tech entrepreneur with early successes like Zip2 and X.com. He was also instrumental in launching the company that later became PayPal.

When eBay bought PayPal for USD$1.5 billion in 2002, Musk netted a USD$180 million mega-fortune from the deal. Musk did not end up relaxing with everything his new millions could buy. In 2002, he founded

SpaceX with the almost ludicrous mission of colonizing Mars. Eventually, Musk's investments began to pay off.

Elon Musk invented Zip2, arguably the first-ever electronic city guide. This was his first software company, and he wrote the software himself. He developed a few new technologies for radius searches and directory management when building this software. Ultimately, this company was sold to Compaq.

Elon Musk is an entrepreneur launching SpaceX, a private aerospace design and manufacturing company. In a different front, Musk contributed USD$30 million to start the company Tesla; his main goal was to produce sustainable electric cars empowered by advanced digital technologies.

Elon Musk does not bend under pressure or from a crisis. He enjoys solving problems, such as actively working to solve existential problems like climate change and traffic gridlock. He has also tried to solve immediate issues like ventilator shortages to varying degrees of success. Here is Elon Musk's superhuman vision: in this future, energy will be inexpensive, sustainable, and abundant; people will work in harmony with AI machines and even merge with them. Humans will become an interplanetary species. While Mr. Musk may not be an expert on everything, he has proved being a champion in the tech and digital business fields.

Chapter 31

From Traditional to Digital Media – Freedom and Its Boundaries

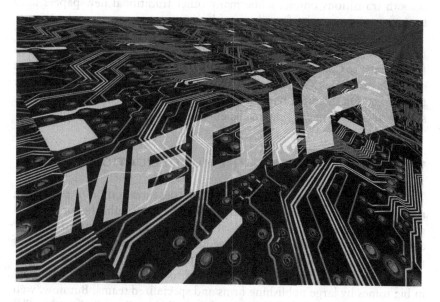

Digital media consists of email, videos, blogs, Twitter, Facebook (Meta), Youtube, LinkedIn, etc.

Digital technologies started with computerization, and then around 1969, when the Internet came into play, it quickly began to impact communications and information sharing. In 1989, a British scientist Tim Berners-Lee invented the World Wide Web (WWW). The web was originally intended and developed to meet the global demand for information sharing between scientists in universities and institutes.

In 1993, Tim Berners-Lee wrote the first version of HTML (HyperText Markup Language), which is the standard markup language for digital documents to be displayed in a web browser. Since then, as described in Chapter 10, the WWW represented the birth of Web 1.0, which allowed people to share, view, and read information online.

DOI: 10.1201/9781003305187-36

The Web also disrupted the traditionally so-called "paper media," newspapers in particular. Compared to traditional newspapers, online digital media has clear advantages: it is faster, cheaper, can be updated at any time, and, most importantly, can reach a much wider audience. In Web 2.0, digital media has even become interactive and self-publishing, getting really popular among the younger generations.

The major challenge is on the commercial part of digital media. A lot of people tend to read everything online for free, so the online subscription models need to work hard to find and secure subscribers who are willing to pay a fee for the information provided. In the United States, renowned newspapers like *Wall Street Journal* and *New York Times* have made their smooth transitions online, while many other traditional newspapers have been struggling or even forced out of business.

Digital media never stop bringing about new changes. Although TV broadcasts rushed to become digital, the Internet has also made over-the-top (OTT) (refer to Chapter 23) possible and allowed the audience to watch videos online directly. This also brought in new competitors like Netflix and Hulu in the market. On the other hand, Twitter, Facebook, YouTube, LinkedIn, etc. started to create a new digital media environment called social media.

THE INNOVATIVE LEAP – THE MULTIPLICATION OF MEDIA

The biggest impact of digital media is about decentralization of publishing rights. Digital media now make everyone can be a self-publisher or personal media, which are run by an individual rather than by an organization. They are generally contrasted with mass media which are produced by teams of people and broadcast to a general population.

Also, dictionaries and encyclopedias used to be assembled and published in big tomes by large publishing firms and specialized teams. But now, web portals like Wikipedia and tons of so-called online dictionaries like Dictionary.com are just a click away online. Online dictionaries offer access through large databases to a word's spelling and meanings, plus ancillary information, including its variant spellings, pronunciation, inflected forms, origin, and derived forms, as well as supplementary notes on matters of interest.

But one may take a second thought in considering any online encyclopedia and dictionary to be 100% trustworthy or credible. Most such online portals only contain the most common definitions or usages associated, and the sources of the definitions can be random and non-authentic when such online versions are open for public editing.

Even worse, some online sources have become the hotbed of hatred, racism, pornography, fascism, and extremism. Even Twitter now has got more content of pornography (source: https://nypost.com/2022/10/26/porn-crypto-more-popular-on-twitter-as-users-lose-interest-in-fashion-celebrities/). Freedom of speech can take a big U-turn here and spark more negative impact to the society and young people. This is the downside of digital media multiplication.

NEW MINDSET FROM DIGITAL MEDIA

Technical Outlook – Optimizing the Internet

How to cope with the Internet and digital media? That is the question. Today, several countries worldwide have blocked or restricted social media, including North Korea, Turkmenistan, China, Russia, Belarus, Iran, and Uganda, to name a few.

Meanwhile, the least censored Internet countries and areas are: Iceland, Estonia, Costa Rica, Canada, Taiwan, the United States, Georgia, Germany, Japan, and Australia (sources: https://pandavpnpro.com/blog/most-least-censored-countries).

The issue is that while authoritarian censorship is clearly not the right way and more politically motivated, the total open Internet and digital media may pose ethical and social issues as well. Just like any routine production ecosystem, the Internet indeed has produced a lot of garbage too. What do we do with the online garbage? They need to be cleaned out from time to time. Then how? How to manage the fine line between freedom of speech and abuse of freedom of speech? This critical issue remains to be addressed effectively by the world.

For a good example, Canada is setting up a good approach to better handling Internet content, and its Canadian Center for Cyber Security (source: https://cyber.gc.ca/en) serves as the single unified source of expert advice, services, support, and guidance, on cyber security for Canadians. While many other countries may have multiple channels with responsibilities to handle online issues, the efficiency and effectiveness are low in the end.

Starting from January 1st, a California law in the United States takes effect and imposes new obligations on social media companies to monitor and potentially regulate the spread of false information online (source: https://news.bloomberglaw.com/us-law-week/what-tech-firms-should-know-about-californias-social-media-law). The Califiornia congress is also working towards a new state bill that could make social media companies liable for how their platforms affect children.

Things can go and stay normal only along proper and rational boundaries. There is no such a thing called ideal mode which is equal to the extreme mode. Do not let the Internet and digital media become a two-bladed sword. We need to uphold the principle of freedom of speech, and meanwhile hold people responsible and accountable for their online presence and behaviors. Do not wait until the dark web causes severe damages to society and younger generations.

Business Outlook – Digital Marketing Heating up

For businesses, digital media opens vast opportunities for online or digital marketing, which is the promotion of brands and products to connect with potential customers online and via other forms of digital communication. This includes email, web-based advertising, social media, text, and multi-media messages as a marketing channel.

Compared to traditional marketing, the main sameness and differences between digital and traditional marketing include the marketing mix and the medium to convey a marketing message. Like traditional marketing, digital marketing also has three main components: lead generation, lead capturing, and lead nurturing. While traditional marketing uses traditional media like magazines and newspapers, digital marketing uses digital media, such as social media or websites. The advantages of digital marketing include global reach, lower cost, saving time in rectifying content, and effective targeting.

A digital marketing strategy is for setting up an online presence through various channels such as organic search, paid ads, social media, and your website so as to achieve target marketing results. There are five Ds as core elements when launching digital marketing: digital devices, digital platforms, digital media, digital data, and digital technology. The retail industry, for instance, is among the very top when it comes to the ROI (return of investment) that you can generate through digital marketing. With more and more consumers looking for speedy online solutions for purchasing products, retail companies have no choice but to take their business online.

Application Outlook – Exploring the Digital Media Horizon

Digital media includes digital photographs, books (ebooks), Websites, and Blogs. Social Media (Facebook and Twitter), Mobile Phone Apps, and emails. Digital marketing campaigns cover content marketing, email marketing, mobile marketing, and marketing analytics. Content marketing is the heart of most successful digital marketing campaigns. All means have to boil down to content which is the catalyst to growing your digital footprint, generating inbound leads, and building the brand recognition you desire.

Typical examples of digital marketing include Social Media Marketing (SMM) – doing marketing via the social media network or using social media apps as a marketing tool. Social media can make sales sound softer and easier to be accepted by the audience. Hence, we can use social media to build a brand; increase sales; drive traffic to a website; Search Engine Optimization (SEO) – the process of getting traffic toward your sales from organic, free, editorial, or natural search results in search engines. SEO will be explained more in the following use cases.

From digital media success stories, e.g., TikTok, we can tell that digital media need to go three ways: conceptually attractive, technology friendly, and user awareness. TikTok has clearly figured out that globally users spend an average of 52 minutes daily on the app, with 90% of users access. TikTok, therefore, decides the most engaging approach for social media apps is a session length of an average of 11 minutes, where users feel at home.

TikTok's main selling point is that it features built-in recording and editing capabilities that make it easy for content creators to record, edit quickly, and post video content on the platform. These short clips can quickly go viral when done right. Plus, TikTok has advertised online everywhere young people hang out – including competitors' social sites, such as Snapchat and YouTube.

INSIGHTFUL PRACTICE

Twitter and Its Spam Accounts

Twitter, founded in March 2006, is an online microblog where users can publish short messages with no more than 280 characters. Today, Twitter is one of the top 3 social media apps in the United States and has around 450 million monthly active users as of 2023 across the globe (source: https://www.demandsage.com/twitter-statistics/#:~:text=Twitter%20has%20around%20450%20million,daily%20active%20users%20(mDAU).

What is behind Twitter's success? First of all, Twitter sets a natural tone and casual communication among users. Unlike other popular social media, Twitter is a conversational social platform where users are eager to engage with the content they like and reshare it, and add their thoughts on the matter.

Twitter's design is kept extremely simple and user-friendly, and you can customize your feed, the topics, and the people you want to keep up with. The platform is straightforward to navigate on both mobile and PC. Twitter makes itself unique because of the real-time conversation users can join. Twitter is also a "go-to" spot if you want to follow influential and VIP-level people.

However, as tricky as any digital media can be, Twitter's fate can face some twists too. In April 2022, Elon Musk announced his plan to buy Twitter for $54.20 per share. But he backed away from the deal just a few weeks later, saying he was concerned that spam accounts on the platform were higher than Twitter had claimed. Although the dispute was eventually resolved and Mr. Musk now is in charge of Twitter, the issues exposed remain big and serious.

What are spam accounts? Spam in social networks refers to the unwanted, malicious, unsolicited content or behavior manifested in various ways, including microblogs, messages, malicious links, fake friends, fraudulent reviews, etc. Spam accounts, if liked and followed, can lead to a mass amount of spam comments and direct messages, and damage the real values of social media.

A new study as of 2022 estimates that over 10% of Twitter active accounts post spam content — double the company's own claims (source: https://variety.com/2022/digital/news/twitter-10-percent-accounts-spam-1235288118/). Meanwhile, Twitter claims it removes over 1 million spam accounts each day.

Twitter could lose users who are concerned, frustrated, or even harmed by spam bots and fraudulent activities. Persistent security issues could also draw more surveillance from regulators over Twitter and the broader tech industry. Welcome to the pros and cons of the digital world.

Marketing via Search Engine Optimization (SEO)

SEO is the online practice of promoting your website to rank higher on a search results page. In this way, you receive more traffic and viewing. The aim of SEO is typically to rank high on the first page of Google search, which means the most to your target audience. The rule of thumb for SEO is: the higher the website is listed, the more people will see it. So your SEO winning strategy simply is to show up first and high in Google search results.

Search Engine Optimization (SEO) means you want your website or portal ranks high in Google or other search engine's result list. The higher in the list, the more your site will be visited and viewed.

SEO typically involves five main processes: Keyword research and usage – find what people search for and use them in your tags and content. Targeted Content creation – craft content suited for potential searchers. On-page SEO: make your content as clear and standing out as possible in the web portal. Weblink building: build trust and authority from other websites. Technical SEO: improving the technical side of your website, like site speed and loading errors. Site structure: the structure of your website shows Google which pages are most important. This means your site's structure with the right content can influence the ranks in the search engines.

SEO makes a very important part of your online marketing strategy because it helps build brand awareness, drive organic traffic to your website, build authority and trust within your industry, and nurture relationships with new and existing audiences. SEO aims to highlight a company's visibility in random online search results, which leads to more customers and more revenue.

You can either pay a digital media marketing to manage SEO for you, or you can do SEO as DIY SEO (Do It Yourself SEO). With sufficient data and lots of practice, you can learn how to do SEO for your business. Again On-page SEO refers to SEO factors and techniques focused on optimizing aspects of your website that are under your control. Off-page SEO refers to SEO factors and strategies focused on promoting your site or brand.

Burger King's Digital Marketing Success

Burger King (BK) is an American-HQed multinational chain of hamburger fast food restaurants. Burger King's menu started from a basic offering of burgers, fries, sodas, and icecreams and now has expanded to a larger and more diverse set of products. It operates nearly 20,000 outlets globally to date and serves 11 million customers globally every day.

Burger King has been innovative in establishing its digital marketing footprint. Their digital campaign, like TV or online ads in the United States, constantly tip-off competition among the two brand rivals, Burger King vs. McDonald's, also often making quite some phenomena in the digital marketing space.

Burger King is a brand that loves to play with new technologies to create a marketing experience that wows its consumers with pleasant surprises. Social media is the biggest ally of Burger King in the digital space, and they use a variety of tactics on social media to gain brand awareness and audience engagement. On Instagram and Twitter, the US-based Burger King handles have over 1.9 million followers.

A famous example of this BK's digital marketing success was the "Whopper Detour" campaign in 2018 which applied a geofencing app that allows marketers to advertise to smartphone users within a specific radius. It is a mix of different technologies, GPS, radio-frequency identification (RFID), Wi-Fi, Bluetooth, etc.

During the campaign, all the mobile app users had to do was to go close to 600 meters of selected McDonald's restaurant locations. Once they get near, they receive an app notification offering a Whopper burger for one cent. Eventually, this campaign generated 1.5 million downloads of the app, and switched many McDonald goers to BK customers.

Chapter 32

From Hierarchic to Digital Culture – Shaping A More Dynamic and Intelligent Society

In the digital era, information is shared globally and teams from all over the world can work closely together.

The subject of this book is digital mindset leaps. The more mindset leaps happen, the more we progress along toward a new digital culture. A note is that during the process of digital transformation (DX), many organizations believe they need to build up a digital business culture internally as the foundation for all the changes. But this is not our topic here.

The digital culture we discuss here is more societal and holistic in scale, and it is a concept that describes how digital transformation is reshaping the way that we interact as humans together with our close partners: machines and devices. It is the way that we act, think, and communicate within society.

Culture can be defined as our ways of life, including customs, values, beliefs, traditions, and norms of a nation or civilization from generation to

DOI: 10.1201/9781003305187-37

generation. Hence, digital culture is the result of technological innovations. It is applicable to multiple topics but boils down to one pivotal theme: the relationship between humans and technology.

A good example of digital culture is the shared experience of online events such as a global virtual business forum. The norms and shared ideas interact in real-time amongst the forum participants, bypassing and transforming geographic, time, and social differences.

A digital culture matters to all of us, but centrally it is because it is backed up by digital transformation and boosts new thinking and best practices in our life and work.

First, digital culture breaks hierarchy and speeds up a flat world. It enhances people's autonomy and society's democracy. People are now empowered with information and tools to make their own judgments and decisions, which means breaking down the social hierarchies. Second, the digital culture encourages innovation. It fosters an environment that motivates people to learn and try new things and skills.

Third, digital culture attracts young talent and boosts new working culture. Youngsters no longer want to work in a 9-5 rigid schedule. They want to be part of the new digital culture and globalization that allow a collaborative and autonomous workplace. It also increases people's social engagement, allowing them to articulate their voices and opinions to help create an impact.

THE INNOVATIVE LEAP – NAVIGATING THROUGH ALL CHANGES

Knowledge is power; now, through the Internet, we can all access reams of information and data within seconds just by searching online. Digital technology has made it possible for us to have instant access to online courses, training, books, journals, publications, and other information sets. The value is obvious.

Automation is multiplying productivity. Digital technologies are contributing to the automation of processes and machinery across industries. This gives people more time to focus on other issue areas but also assures higher safety standards, protecting us from risk in construction, mining, and other physical industries. Moreover, it has cut the costs of certain tasks by enabling us all to save money via direct access to the final product or service.

NEW MINDSET FROM DIGITAL CULTURE

Technical Outlook – The Digital Environment

A digital environment consists of core technologies and new concepts. The top eight digital transformation technologies, as already covered in this book, include mobile services like our smartphones, cloud services like AWS and Azure, IoT apps like smart home and smart city; broadband Internet access, robotics in manufacturing and other fields, artificial intelligence and machine learning for automation, and augmented reality for Metaverse experiences.

There are emerging new concepts in parallel with the advanced digital technologies that include:

- customer experience (CX)-centric vs. sales and marketing-centric;
- data first or data-driven vs. impulsive decision-making;
- risk-taking and innovative vs. conservative operating;
- collaborative and open source vs. unilateral and proprietary closure;
- agile and scalable vs. fixed and cumbersome;
- instant and transparent vs. delayed and hidden;
- lifelong learning and training vs. regular education period.

Digital transformation will mainly impact the following four areas, each with vast opportunities together with some challenges: work process, business model, technical domain, and organizational culture. There is no one solution fits all approach for process transformation, and digital technologies will enable agility and scalability. Business model transformation is surely no stranger: from physical store shopping to eCommerce, from wired phones to mobile phones.

Next is domain transformation, where we move an on-prem business to the cloud and move the hardware to software-defined virtualization. Lastly, cultural transformation as our beliefs, values, customs, and norms will change with the digital trends, including the focus of this book – our mindset is making leaps during digital transformation.

Business Outlook – Digital Way of Thinking

As quoted in the Preface of this book, based on *Harvard Business Review* published in May 2022, "A digital mindset is a set of attitudes and behaviors that enable people and organizations to see how data, algorithms, and AI open up new possibilities and to chart a path for success in a business landscape increasingly dominated by data-intensive and intelligent technologies" (source: https://hbr.org/2022/05/developing-a-digital-mindset).

Here, we can highlight some key points in this digital mindset definition. First, "A digital mindset is a set of attitudes and behaviors … ." This means that mindset does not stop at values and beliefs; it also comes down to actions; people do things based on what they reason, believe, or perceive. Hence mindset is powerful enough to shape behaviors.

Second, "… that enable people and organizations to see how data, algorithms, and AI open up new possibilities … ." refers to the new digital way of thinking about having things done. The digital way of thinking can be revolutionary and disruptive. For example, to expand a bank's business, the traditional way of thinking is to open up more branch outlets and hire more tellers and advisors. But the digital way of thinking is to move physical banks online via FinTech (refer to Chapter 27), which enables faster go-to-market, more cost savings, and larger service coverage.

Third, "... to chart a path for success in a business landscape increasingly dominated by data-intensive and intelligent technologies." has two folded meanings. First, "to chart a path for success" indicates leadership, vision, and roadmap for digital transformation. Second, "... data-intensive and intelligent technologies" indicates we must know how to leverage digital technologies in every step or phase of digital transformation and use the technologies not just as supporting tools but as pivotal roles to revamp the whole business models and architecture.

Application Outlook – Welcome to More Winning Apps

In this digital explosion era, we see tons of digital applications (apps) available for us to download and use from Apps or Play Stores. But do you know that out of these many digital apps developed and available, only 0.5% of apps are successful, and below 1,000 downloads count for 67.8%? (source: https://www.zippia.com/answers/what-percentage-of-apps-are-successful/).

That tells us how competitive this software app market is, and it is a very tough task to develop and market a successful app. We can check into this from two angles. A good app should make a combination of both user and business-centric elements. It creates value, provides a delightful experience, and great performance for users. While for businesses, it encourages user retention and generates revenue.

First from the users' angles – for winning digital apps, they need at least to carry five key characteristics: great UI (User Interface): the app must meet what people perceive as a great app, and first impressions are everything. Fast loading time and high performance: this may sound like a no-brainer, but if not, the app will quickly be ignored or uninstalled. Quick adaptation to users' needs: this means the user can see that the app can help them with their needs.

Then from the app developer or software company's angle, the idea of the app should come first because behind every great app is a great idea. For example, the Uber's app enables the whole ride-hailing business model. Second, identifies a target demographic, meaning you know the target market you want to serve.

Beautiful UI design attracts user trial and engagement. Frictionless and simple navigation feature can get the users up to speed. Responsive means technical excellence, and the app works seamlessly and responds quickly via all APIs. Lastly, regarding security, the app should guard the users' data and privacy without flaws; in case not, that may lead to immediate failure of the app.

For software firms, the most important KPI (Key Performance Index) for digital application performance includes load time, crash reports, and device information with screen resolution and operating systems. These metrics allow developers to control the app's technical performance, improve the testing process, and enhance user experience.

INSIGHTFUL PRACTICE

Digital Transformation Prevailing Over COVID-19

In the midst of the global pandemic starting from early 2020, digital technologies and apps have stood up high and shine for their capabilities to support us in the fight against COVID-19. Before the current COVID that has caused health, education, social and economic crisis globally, advanced digital technologies, such as artificial intelligence, have not put much of their focus on public health management and alternative work and study models.

Now, COVID is pushing the strength of digital technologies to go beyond small circles and to the larger public. Emerging technologies such as AI/ML help expedite vaccine development, choose effective public health measures, and keep the public updated with scientific information. They have also enabled us to move much of our lives and work online, maintaining business, economic, and education system operations when most people are staying home. Digital technologies also help us to remain connected to one another in society and support any medical and life emergencies when they happen.

New efforts to develop and deploy AI/ML-enabled mobile app solutions to mitigate COVID-19 are well underway, and these new tools also respect fundamental rights, including data protection and privacy. For instance, such a counter-COVID digital app will empower individuals to monitor their own health, e.g., alerting them if they have closer paths with an infected individual, provide them with a just-in-time evaluation of their exposure to COVID-19, and facilitate easy customized access to information and medical advice, while maintaining the solid standards of data and privacy protection.

In 2018, the global investment on digital transformation was approximately $1 trillion. In 2022, spending on digital transformation (DX) is projected to reach 1.6 trillion U.S. dollars. By 2026, global digital transformation spending is forecast to reach 3.4 trillion U.S. In fact, 79% of companies acknowledge that COVID-19 increased the budget for digital transformation (source: https://www.progress.com/docs/default-source/default-document-library/landing-pages/dach/ebook_digitaltransformation_final.pdf).

As we are prevailing oCOVID globally, digital technologies have been our best partners in fighting against the virus. This boosts our confidence that the future remains hopeful and promising and that human intelligence and creativity can better control our destinies.

Starlink Supports Ukraine During a Hard Time

We introduced Starlink in Chapter 5, but its big impact wouldn't stop there. Again Starlink, offered by SpaceX, founded by Elon Musk, is a network made of over 2,000 satellites orbiting the Earth, connecting thousands of terminals on the ground. Starlink was activated across Ukraine in late February when Russia's invasion disrupted Internet services.

Starlink's presence in Ukraine now serves as a vital communication line in war-torn Ukraine, complementing precarious regular telecommunication infrastructures. It has proven invaluable both on the combat front and in restoring energy and communication infrastructure targeted by Russia. As the end of 2022, there were 25,000 Starlink terminals in Ukraine.

One direct resistance support example has been the use of unmanned aerial drones. When the drones are connected via a Starlink satellite, Ukrainian artillery teams can target Russian positions by dropping anti-tank munitions with great precision and success.

Starlink was originally envisioned to offer Internet connectivity via satellite to areas with a lack of communications and Internet infrastructure, such as at sea, in mountain areas, and other isolated locations far from cities, or in places with governmental restrictions on Internet access.

Now Starlink has also become a lifeline as Ukraine battles Russian invasion forces. As Russian strikes have caused damage to Ukrainian villages, towns, and cities knocking off the power, Starlink terminals are now vital to Ukraine's ability to connect to the outside world.

Amazingly, Russian forces have attempted to obstruct Starlink Internet connectivity from space by using jammers. According to SpaceX, a software upgrade has been released for Starlink that can avoid jamming transmitters and save power usage. In this sense, Starlink is also directly fighting against the Russian invasion of Ukraine.

Customer Experiences and Digital First

In the digital era, one thing is clear – it is the customer who is in the driver's seat. In essence, experience (CX) today will define and decide business growth or survival. It is about changing the way businesses engage with their customers and how they provide a consistent experience of satisfaction whenever and wherever the customers need it.

Based on the recent Deloitte's report, companies with higher digital transformation rates and maturity reported 45% revenue growth. In total, 29% of highly digitized companies reported strong growth and innovation impact, while 41% resulted in a positive lift on sales and marketing functions (source: https://www2.deloitte.com/us/en/insights/topics/digital-transformation/digital-transformation-survey.html).

This is saying that to deliver a better CX, keep in mind that this is the next-generation digitally savvy and conscious customer. Digital technology has transformed a lot of consumer habits. For purchasing means alone, mobile devices, apps, automation, AI/ML, and much more means allow customers to get what they want on demand in real-time.

Even further, these new digital technologies have shifted customer expectations, creating a new kind of modern buyer. Today's consumers, especially younger ones, are not tied to a single channel. They are app-native, constantly connected, and aware of what they can do with technology.

As a result, customers often rate vendors on their digital CX first. This pretty much means today's CX is a digital CX or digitally driven CX. This requires you as a business to put up a new digital strategy called Digital First, rethink how you engage and interact with your customers, and accordingly come up with a digital sound strategy and implement it quickly and properly.

For instance, personalizing CX is a very effective new approach. Today's buyers appreciate that organizations treat them as unique individuals and serve their needs based on their personal preferences and purchase history. Of course, there is always a fine line between using customer data and abuse of customer data, but as long as you handle customer data properly and based on the rules, the CX will improve, leading to your sales and business growth.

In summary, there are three pivotal components in creating an effective digital strategy for CX: discovery – making it easier for customers to find the merchandises they need; engagement – making it easier and more friendly the ordering and purchase; and delivery – shipping and handling the goods over to the customer's hands. You need to synch up all the digital channels you have in your company in order to provide a single, user-friendly CX.

The digital culture connects more people and countries around the world for better lives.

Afterword – Some Bold Predictions on the Post-Digital Future

I hope this book has helped you go through an exciting and enlightening journey of the 4th Industrial Revolution, better positioning your mindset open and poised for digital transformation. The world has witnessed the growth of a New Economy and Culture at the start of the 21st century. Information in all its forms is getting digital and pivotal in this new era. Data are reduced to bits stored in computers and flowing through networks at lightning speed. Generative AI already starts to create new contents and notions on our behalf.

While we are still developing and adapting to the current Digital Age, at the end of this book, it may help to post the question: what is the next? What will the post-digital era look like? From a holistic and futuristic perspective, we will see digitalization moving along two major fronts: on the one hand is the continuous improvement and upgrade of core digital technologies in terms of processing power, capacity, speed, intelligence, algorithms, cybersecurity, etc. On the other hand, we will have more cutting-edge digital applications, collaboration, and synergy with vertical fields such as biology, manufacturing, healthcare, finance, government, social media, military defense, transportation, education, etc.

It is worth noting that the latter development is not just about new digital applications for those industry verticals or social segments. Instead, with the more profound convergence of digital and other advanced technologies in each field, for example, advanced DNA analytics and reengineering in biology, we may see some epic and brand-new scientific breakthroughs on the horizon.

Digital technologies have advanced more rapidly than ever in any innovation in human history and can be great equalizers by enhancing connectivity, access to trade, financial inclusion, and public services. Digital transformation has reached around 50% of the developing world's population in only 20 years and transforming our societies. This transformation will continue to accelerate in days to come.

If we look toward the future, distributed ledger technology, blockchain or Web 3.0, artificial intelligence, machine learning, deep learning, augmented

DOI: 10.1201/9781003305187-38

and extended reality, and quantum computing will be the next fronts of new post-digital technologies to spark changes. These new technologies will let businesses reimagine entire industries and markets. Technology-driven cultures are creating an expanding technology identity for every consumer.

Technology will keep changing how people learn, think, and communicate and play a major role in society. Now we nearly cannot live without technology. Both modern technology and society are co-related, co-dependent, and co-influencing with each other. It is expected that around 2030, people could start using AI powered robots to work around their houses and provide companionship.

According to Forbes, IoT technology will be embedded in 95% of electronics as new products by 2050. We will have everything connected to the cloud and the Internet (source: https://www.forbes.com/sites/forbesbusiness deliverdevelopmentcouncil/2020/09/24/three-ways-underappreciated-iot-is-ing-for-businesses-quietly/?sh=72ce29c1298f).

Amid all the new technology breakthroughs, recently AI is becoming the hottest subject of attention. While we are moving quickly from the current AGI towaed ASI stage (Refer to Chapter 6), the excitement comes together with risks and concerns. Some scientists are worried that AI might become the very first technology that is invented by humans but losing humans' control. We therefore are seeing more calls to regulate AI development and apps before too late. Any new technology may accompany some new ethical challenges. Let's face them boldly and resolve them with wisdom.

For businesses, the next wave of technology will make it possible to smart companies to emerge. Smart companies automate the production process with AiOPs (artificial intelligence for IT operations), customize products and services, and deliver them instantly on demand. Such businesses can handle the production or service life cycle based on a clear understanding of the needs of consumers, employees, and business partners better than ever before.

Such smart companies will have the agility and creativity to differentiate themselves from go-to-market all the way to customer services, to new products R&D, and make each client feel at home. These businesses will be able to cater to individual clientss in every aspect of their lives, careers, or business relationships and also shape their future requests and desires.

When a client calls for their ICT system upgrading, smart companies can quickly have many custom-designed, software-defined, AIoP-powered modules stitched and tuned together and launch into a holistic and robust ecosystem where the client's business capability and productivity will be lifted to the next level.

Overall, differentiation comes from applying digital technology in powerful new ways in the post-digital world. The technologies must differentiate beyond digital tools and conceptual-level apps and bring about holistic and disruptive innovations and breakthroughs. If you want to grow and succeed and build trust with customers, employees, business partners, and communities, then forging an advanced and responsible synergy with new technologies is a must. This is the era in which humans and technologies work together along the new horizon and make miracles happen where only the sky is the limit.

David W. Wang
Oak Hill, Northern Virginia
May 2023

Index

Printed in the United States
by Baker & Taylor Publisher Services